Culture an

CULTURE AND COGNITION

The Boundaries of Literary and Scientific Inquiry

Ronald Schleifer
Robert Con Davis
Nancy Mergler

Cornell University Press
Ithaca and London

Open access edition funded by the National Endowment for the Humanities/Andrew W. Mellon Foundation Humanities Open Book Program.

Cornell University gratefully acknowledges a grant from the Andrew W. Mellon Foundation that aided in bringing this book to publication.

Second paperback printing 2019

ISBN 978-0-8014-2632-2 (cloth: alk. paper)
ISBN 978-1-5017-4672-7 (pbk.: alk. paper)
ISBN 978-1-5017-3852-4 (pdf)
ISBN 978-1-5017-4673-4 (epub/mobi)

Librarians: A CIP catalog record for this book is available from the Library of Congress

Cover illustration: *Haunted Tree, Dartmoor* by Carol Betsch.

This book is dedicated to our parents:

Cy Schleifer
the memory of Helen Szozkida Schleifer
Edna Ingram Schleifer
Elva Covene Anderson
the memory of Robert Undiano
Hugh G. Davis
Walter Mergler
Janet Whitmore Mergler

Contents

Culture and Cognition

Preface

Like the Copernican revolution, the renovations of science in the twentieth century have wrought changes in the basic conceptions and, as Steve Woolgar says, "the very idea" of science. From Newtonian physics to quantum mechanics, from early twentieth-century sublime versions of science to recent chaos theory, these shifts have brought the "very idea" to bear on every realm of modern culture. As a channel to subcellular and subatomic worlds, as a creator of high technology in the postindustrial world, and as a perspective on the macrosphere and the origins potentially of "everything," science at times seems unassailable and beyond external commentary. This prestige has led many in adjacent but nonscientific areas of knowledge to assume the mantle of scientific authority in the study of the arts, humanities, and social sciences.

But notice our reference to "the very idea" of science rather than "science itself." In this book, we deliberately speak of configurations of a certain kind of discourse and are not presupposing incontrovertible scientific perception, experimental validity, or simple knowing. Our focus is discourse rather than gnosis, and our critical preference for thinking about the *discourse* of science is, of course, strategic and draws us in certain directions taken since Thomas Kuhn's *Structure of Scientific Revolutions* by many in the philosophy and history of science, by social scientists such as Bruno Latour and the Edinburgh theorists, and by scholars who have begun work in literature and science. We are among those, in short, who do not take science to be unquestionable as an institution or beyond commentary in its achievements. Our preference for viewing science as a practice and a class of cultural discourse in dialogue with other cultural discourses says that science is not an absolute grounding or reflection of perception and truth but a dominant discourse. "Scientific experiment," Stanley Aronowitz writes, "may be shown to derive from a specific conception of 'value,' that of intervention into nature as the road to reliable knowledge" (1988: 346). We find this idea of science-as-practice to be productive and take science not as the measure of all conceptions of accuracy and truth but as a version of meaning with instituted practices and a potential for intervention guided by specific values.

Taken as a discourse imbued with value, in other words, science is not pure metacommentary unrelated to other discourses such as philosophy and aesthetics. It was at least conceivable before Michel Foucault's archaeology of knowledge and Kuhn's study of science to maintain the separation between science and culture, the idea of "two cultures"; after Foucault and Kuhn, we envision large discursive patterns in society that engender scientific *and* aesthetic modes of thought and representation. Aronowitz is correct in saying that "the distinction between philosophy, long viewed as a speculative inquiry, and natural science, in which speculation is strictly limited by scientific method to preexperimental hypothesis, has become increasingly blurred" (1988: 347). Instead of two cultures, there is now a stronger paradigm in which social discourses create the potential for scientific and humanistic formations as well as openings for intervention. Aronowitz's historical judgment is that "science is the discourse of the late capitalist and the 'socialist' state" (1988: 352), and he claims that science as praxis and as a set of institutions is perfectly deducible from an economic mode (late capitalism) existing at a particular moment of Western history. His strong theory potentially accounts for the "blurring" of scientific and humanistic discourses in a dialectical articulation of historical events and cultural formations. We are not advancing Aronowitz's Marxist conclusions, and yet we agree with him in principle about the obligation to read scientific discourse as an *interested* practice, an activity constructed to achieve particular social ends and to foreclose others at a particular moment in history.

We are aware that our view of science-as-discourse coincides with the ethnomethodological definition of the "sociology of scientific knowledge" and, at least in theoretical orientation, with such works as Latour and Woolgar's *Laboratory Life* (1986) and Latour's *Science in Action* (1987). In *Science: The Very Idea* (1988), Woolgar describes the perspective of the "sociology of scientific knowledge" (SSK) as deriving from "a range of disciplinary interests in science: notably sociology and history of science, less prominently, philosophy, anthropology and psychology" (1988: 14). Underlying this interdisciplinary amalgam is the notion of social "discourse," the assumption of isomorphic units in a system of exchange, suggesting a theory of culture based on the signifying function as an instrument for the engagement with culture as a multiplicity of discourses. A main theoretical tenet of SSK, rejecting the status of science as a unique instrument for attaining empirical accuracy and truth, is the theoretical positing of "historical and cultural relativism" (1988: 14), the absence of a last or ultimate frame in which to place "true" science or from which to orient the unobstructed and fully accurate scientific view. Such relativism, we believe, however, dictates not chaos and incoherence, but the persistent complexity, as Woolgar also notes, of needing to define "meaning" locally and *in situ*, as the sum of "language (representation) + context" (1988: 57). This paradigm of knowledge construction is at once relativistic and rigorous in its aim always to situate knowledge materially and historically. It aims to *redefine* the referential aspects of language and scientific understanding.

The specific situation of this book is our staging of the encounter of semiotics, cognitive science, and psychoanalysis—the superimposing of these discourses within the same interpretive context that we have constructed for studying scientific and humanistic discourse. Of course, we have had to confront fundamental and difficult questions. How can we speak of cognitive science in relation to semiotics? How can we superimpose cognitive science and psychoanalysis within the frame of semiotics? How does the deliberately reductive and simplifying function of cognitive modeling relate to the familiar "comfort" of ordinary experience as discussed by feminist theorists of science or the "comfort" effected by the functional repression of theory and conflict? In staging these dramatic encounters among semiotics, cognitive science, and psychoanalysis, we are exploring further questions about the nature of critique and about the construction of a cultural discourse within which to articulate the relationship of scientific accuracy *and* humanistic comprehension—truth and meaning, knowing and understanding. The three authors of this book do not explore these questions separately as a semiotician, a cognitive psychologist, and a Freudian—although that scenario is not *entirely* wrong; rather, we are all three trying to discover the points of intersection among discourses that, in fact, interact within various cultural paradigms.

Exactly *how* to write this book is the problem we faced at every stage. How do we three begin to speak to one another about such different discourses? Where does the encounter of semiotics, cognitive science, and psychoanalysis begin? Does it develop or progress as an encounter? Does this staged production have a theme, a plot? How will this encounter end?

Can it end? In the Introduction, we begin to orient these questions in increasingly complex interpretive relations. In terms that we return to repeatedly, we examine the *simple* binary differences that make up mathematical relations as well as the rudimentary couplings of semiotics, one as opposed to two, black as opposed to white, and so on. The simple distinction, like Ferdinand de Saussure's early semiotic descriptions of cultural institutions, posits a system of exchange among various isomorphic cultural practices, the substitution of signifiers in some practices for those in others, and the occlusion of signifying possibilities as others are promoted into prominence. We also examine the distinctions of science that attempt to be *exhaustive*. By exhaustive understanding, we mean the attempt to survey and accurately monitor a whole field of inquiry that generally goes under the heading of scientific empiricism and is expressed in the practice of cognitive science. And, finally, we look at the *generalizing* descriptions that project *discourses* themselves as the objects of inquiry. This approach attempts to predict the appearance or articulations of phenomena, the ability to account in advance for what does not yet exist, for the yet-to-be-observed.

We have few illusions about the prospects for creating a fully successful, strong model that will work equally well on all accounts for science and larger cultural representations. Our goal, nonetheless, is precisely to articulate semiotics, cognitive science, and psychoanalysis as complex relations in a discursive scheme. This large strategy shapes the three sections of this book. In the Introduction, we discuss the concepts of simple, exhaustive, and generalizing explanation in a scheme of gradually increasing complexity. Our aim there is to test and explore our three-part scheme within ongoing debates in the philosophy of science and, further, to situate the discussion of our book within conceptions of culture and cultural studies. A major theme within our three-part scheme is a working conception of narrative cognition. This conception is close to Louis Mink's definition of narrative comprehension as "grasping together in a single mental act things that are not experienced together" (1970: 457). Throughout *Culture and Cognition* we repeatedly return to narrative structures and activities in examining the claims of cognitive science and situating those claims within the larger domain of culture.

In Part I, "Narrative Structures," we move to substantiate our presentation not so much theoretically but in three actual instances of inquiry investigated according to the model we are proposing. In Chapter 1, we discuss attempts to account for cognitive science as a discipline in relation to adjacent (and simple) binary schemes in linguistics, semiotics, and discourse theory. We show the degree to which semiotics presupposes operational (cognitive) axioms concerning the existence and function of signs. We also discuss the degree to which cognitive science builds itself more or less unconsciously out of semiotic narrative constructs—having to do with cause-and-effect relations, normative bases for proofs, and research-based models for exhaustive documentation—that are often taken by scientists to be invariant features of a research model.

In Chapter 2, we construct a theory of literary genres of narrative according to the rigorous semiotic view of discourse elaborated by A. J. Greimas. We explore the cognitive implications of constituting literary genres according to semiotic relationships, that is, in terms of generalizing and even predictive descriptions of genres. This is a multilayered discussion, and we intend its complexity to demonstrate the situating of a set of phenomena as at once simple, exhaustive, and generalizing. We intend this example to show that all three levels of this typology are present in any involved inquiry focusing on cognitive activity—even one, such as Greimas's, that rigorously aims at simple *generalizations*. In Chapter 3, we examine the attempts of cognitive scientists to be "exhaustive" and "accurate" in their study of the storytelling and narrative practices of old people. Their task involves establishing categories for empirical inspection, such as the response to "noise" in the environment, the age of the interlocutor, and the complexity of information being communicated—all categories that constitute a potentially comprehensive cognitive mapping of the field old people respond to as they narrate their experience. The attempt here is to describe dimensions of that cognitive map in an accurate and verifiable—that is, reiterable—manner and, at the same time, to examine the cognitive-narrative strategies of "natural history" in order to present a critique of the simplifications of *exhaustive* modeling of cognition. In fact, the chapters of Part I together aim at presenting and critiquing the idea of simplifying models of cognition.

The complexity of the examples in Part I is indicative of the multiple relationships of culture and cognition, and in the further developments of the book we foreground not the separation but the complex intersection of the three phases of our scheme in each instance of inquiry. In Part II, "Cases of Cognition," we turn more fully to the configurations of science, cognition, and discourse. In Chapter 4, we explore the examples of Sigmund Freud and Albert Einstein as "special cases" of scientific projects that are informed by implicit orders of cultural representation. These are narrative orders, Freud's conception of Oedipus, and Einstein's special and general relativity as narrative accounts of signifying practices. In

Chapter 5, we explore the broad implications of psychoanalysis as a narrative theory of cognition and explicitly stage an encounter between Jacques Lacan's semiotic version of psychoanalysis and Greimas's theory of cultural discourse. In effect, we superimpose Lacanian psychoanalysis on the square of Greimassian semiotic and cultural theory. In Chapter 6, we look once again at the language of old people, this time focusing less on methods of investigation and more on the particular linguistic and extralinguistic factors that shape old people's language as understood in the investigations of cognitive science. Again we are exploring discourse within semiotically informed narrative and cultural theory.

In Part III, "Cultural Discourse," we attempt to move beyond the confinements of explicitly scientific and humanistic inquiries into a cultural critique that assumes the relationship between those previously disparate discourses. We examine approaches to criticism and pedagogy actually practiced in the academic institution of the American English department and in the institution of professional journal editing. Acknowledging but no longer seeking to separate simple, exhaustive, and generalizing distinctions, we here focus on the possibility of critical discourse as an *interested* cultural criticism. This possibility begins to move criticism out of exclusive confinement in the academy and positions it as a kind of cultural activity, just as in Chapter 3 we examined the ways cognitive science positions cognition as a social activity. The situating of literary studies as a cultural activity draws on both semiotic formulation and scientific attention to testability and accuracy in actual situations—all seen as the critique of practices and aims implicit in a particular inquiry. In our discussion of pedagogy and professional publishing, we attempt to situate ourselves not as either scientists or humanists but as critical, *interested* investigators—oriented critically and scientifically but acknowledging our own role *as investigators* who are also imbricated within the field of inquiry. In an important sense, the purpose of our discourse in this book is to align ourselves so as to be able to speak with the particular voice of cultural discourse in Part III, critical and interested but neither merely scientific nor merely humanistic. We are attempting here a kind of "natural history" of a particular institution of cognition where the emphasis is on *history*—and the *emergence* of understanding we describe in the long single chapter of Part III—but in which, as in Greg Myers's description of natural history, the role of the observer plays a prominent part so that cognitive activity itself can be seen to be narrated and situated (1990: 201–3). For this reason, the "cultural discourse" of Part III is an extended, multifaceted chapter rather than resembling the three-chapter structures of Parts I and

II. Among other things, it examines the problematic relationship between general and special cases of cognition by examining ethics. (That problematic relationship is inscribed in Greimas's semiotic square, which we use throughout *Culture and Cognition* in analyzing both logical abstractions and particular semantically charged designations. We distinguish between the two by italicizing the latter.)

The principal theme of the disciplinary encounters of this book is the movement from simple critical binarity, through elaborated cognitive and narrated framings of binarity, and, finally, to a cultural critique that carries with it, as technologies and strategies of positioning and decipherment, the structures of humanistic, scientific, and narrative discourse. In the process of this movement, the semiotician, cognitive scientist, and psychoanalyst can begin to speak to one another's disciplinary and interdisciplinary interests. They do this through a discourse that relies on and deploys axiomatic and relatively unself-conscious positing of the objects of investigation and research, all the while reserving and then advancing the active critique of the grounding of investigation. This is discourse guided, as Greimas advances, not just by the tropes of *either/or* and *and* but also by the cultural activity of the negative complexly conceived, *neither/nor*. Thinkers as diverse as Kenneth Burke, George Steiner, Shoshana Felman, Umberto Eco, Bruno Latour, Julia Kristeva, Theodor Adorno, and Katherine Hayles have argued in different ways that negation is a significant instance of cognition and cultural activity. It is conceived in terms of "contrary to fact" and "potentiality," and it is perhaps even the motor of cognitive and cultural activity. It leads both to the institutions of understanding and to their critique, and in studying culture and cognition it is important not to lose sight of either function.

Our book's theme of the relationship of binarity and cultural critique is made evident in a further way as well. We have foregrounded three exemplary theorists in our book—Darwin, Greimas, and Lacan—precisely because their work emphasizes the intertextuality of scientific and humanistic discourses. In addition to these three, we could easily be discussing Claude Lévi-Strauss, Julia Kristeva, Jerome Bruner, and others as well. However, we have given discursive priority to Darwin, Greimas, and Lacan, not always with explicit acknowledgment but, we hope, with evident consistency and effect. In Darwin we explore the simplicity of binary couplings and the complications of narrative "histories" evident in his theory of adaptation. He is also important to our discussion for the generalizing and predictive features of his naturalistic economy of explanation—an economy that fosters the questioning of the self-evident, as evidenced in the title of Chapter 3, "Why Are There Old People?" Through Greimas, who pursues logical rationalism the most rigorously of the three, we explore the semiotic modeling of cultural inquiry and discourse and attempt to expose its implicit theory of cognition. We also make extensive use of Greimas's "semiotic square" as a *semantic* modeling of social and cultural relations.

In Lacan we attempt to move in the interstitial space between science and culture, explicitly to articulate the relationship of culture and cognition. Lacan's own typology of Imaginary, Real, Symbolic, and Symptomatic orders is a strong model for the scheme we advance in this book. (In Chapter 5, we inscribe these orders on a semiotic square.) Like "simple" distinctions, the Imaginary order foregrounds the relations of *either/or* in formulations of logical exclusion. That is, the Imaginary order is made up of binary and largely privative relations between presence and absence, yes and no, on and off, and so on, formalistic stagings of positive and negative relational possibilities. Against the simplification of the Imaginary, the Symbolic presents generalizing distinctions suggesting that information is almost totally connected and cross-referenced, totally patterned. In other words, Lacan's Symbolic register (borrowed from Lévi-Strauss, and in many ways homologous with the symbolic order of Greimas's analyses) suggests patterned distinctions and recursions in language and cultural discourse, deployments of imaginary relations and seemingly "real" facts in phases and cycles calculated to accomplish particular aims and satisfy certain desires. Lacan's third order, that of the Symptom, corresponds roughly to simple (and simplifying) empiricism, binary-symbolic meanings mistaken as empirical truths (Lacan's "Symptom," finally, is much more complicated than this). In Chapter 3, we cite discussions of adaptation that pursue such simplifying empiricism so single-mindedly—in isolation from the semantic and cultural values of the terms they traffic in—that they approach unconscious parodies of Darwin. The "reality" of empiricism, as Lacan defines it under the term "Symptom," is actually the "impossible" phantasm of monologic and isolated meanings, the phantasmic idea of a pure showing forth of information prior to interpretation or theoretical framing. For this reason, Lacan's order of the Real functions as a critique of the prospect of purely empirical disclosure. It is the contrary to the "empirical" reality of meaningful "symptoms" and foregrounds the impossibility of essential and nonrelational instances of information. As such, in its very inarticulability, it does the work of the negative we mentioned above, situating the orders of the Imaginary, the Symbolic, and the Symptom as always *emergent* categories and institutions of understanding.

In short, we foreground Darwin's, Greimas's, and Lacan's discourses as key textual references for the typology we are advancing in the articulation of culture and cognition. We intend all these discursive strategies to contribute to the construction of a critical discourse for the articulation of science and signification (representation) within the frame of cultural studies. By using the term "cultural studies"—suggestive of the Birmingham Centre, interdisciplinarity, and transnational studies (see Brantlinger 1990)—we intend neither a scientizing (sanitizing?) of the humanities nor a humanizing (humoring?) of the sciences. Our intention is not to alter or reimagine these discourses, even if we could, or to dislocate them from the cultural references and function that give them power. We attempt to focus on the ways in which they *already* participate in the construction of an instituted cultural economy. In the manner of cultural studies as it is emerging as a disciplinary institution in the United States, we are seeking, as one of our principal aims, to substantiate the case for seeing that scientific and humanistic discourses *are* practices with social agendas and commitments to cultural values, values that frequently do not correspond to the self-descriptions offered by those practices.

We also intend the sequence of our discussions in this book to narrate a version of the terrain of modern cultural theory. In this broad narrative, in the first part of the book, "Narrative Structures," we discuss aspects of Anglo-American philosophy and the advent of semiotic (binarist) and structuralist paradigms associated with modernism and early twentieth-century social science and linguistics. Within Part I we superimpose cognitive science, empiricism, and the narrative power of natural history. Our aim here is to frame the simplifying and exhaustive gestures of empiricism—to situate empiricism as a social practice. In the next part of our narrative, "Cases of Cognition," we examine discourse in Freud, Einstein, Lacan, and Greimas, theorists whose discourses are typical of mid-twentieth-century attempts to theorize psychological and worldly relations in complex amalgamations of simple, exhaustive, and generalizing economies. At the end of this section we describe the rhetoric of narrative in terms influenced by Jacques Derrida. In all these complex discourses, we attempt to discern narrative formations that are neither purely logical nor simply accidental. In the last part of our narrative, "Cultural Discourse," we examine institutions of cultural and social discourse in a Foucauldian frame. These are discourses that emphasize power relations and social institutions in a multifaceted discussion of ideology, pedagogy, and scholarly publishing. In our book's large narrative mapping of intellectual terrain in the twentieth century, a central theme is that of the rise of cultural and social theory. What results is a paradigm that creates an emergent understanding of the interrelations, and superimpositions, of scientific and humanistic culture.

A book such as this, which aims at configurations of interdisciplinary understanding, perhaps necessitates its authors working more closely with other scholars than do more conventional studies. *Culture and Cognition* is most indebted to James Comas, Michael Goldstein, and Alan Velie. Each of these scholars collaborated with us in work that, with much revision, has been incorporated in this book. Their thought and, in some cases, versions of their sentences have found their way into the arguments reconfigured here. We have greatly benefited from working with them. Discussions with and readings by many other scholars have also been very important to our work. The readers for Cornell University Press made invaluable detailed comments, and many of the strengths of our project have been a result of their care. The comments and conversation of friends and colleagues here at Oklahoma—David Gross, Hunter Cadzow, Susan Green, Frank Durso, Richard Barney, and Monica Gregory—have substantially contributed to our thinking and argument. In addition, the support of Provost Joan Wadlow, Dean Rufus Fears, and Bernhard Kendler of Cornell University Press has eased and facilitated our project. Finally, Peggy Frazier and Steven B. Wilson gave us important help at a late stage of our work. Melanie Wright compiled the index and, as ever, aided us in innumerable ways.

As we mentioned, portions of *Culture and Cognition* began in a number of articles we have published in a wide range of venues. This work has been substantially and, in several cases, almost entirely reconceived and rewritten in a version of the "reconfigurations" of cognition we examine here. Still, we thank the editors of journals and publishers listed below for permission to rework and reproduce parts of the following essays: Nancy Mergler and Michael Goldstein, "Why Are There Old People: Senescence as Biological and Cultural Preparedness for the Transmission of Information," *Human Development* 26 (1983), 72–90; Robert Con Davis, "Introduction: Lacan and Narration," *MLN* 98 (1983), 848–59; Ronald Schleifer, "The Space and Dialogue of Desire: Lacan, Greimas, and Narrative Temporality," *MLN* 98 (1983), 871–90; Nancy Mergler and Ronald Schleifer, "The Plain Sense of Things: Violence and the Discourse of the Aged," *Semiotica* 54 (1985), 177–99; Ronald Schleifer and Alan Velie, "Genre and Structure: Toward an Actantial Typology of Narrative Genres and Modes," *MLN* 102 (1987), 1123–50; Robert Con Davis, "Theorizing Opposition: Aristotle, Greimas, Jameson, and Said," *L'Esprit Createur* 27,

2 (1987), 5–18; Ronald Schleifer and James Comas, "The Ethics of Publishing," *The Eighteenth Century: Theory and Interpretation* 29 (1988), 57–69; Robert Con Davis, "A Manifesto for Oppositional Pedagogy: Freire, Merod, Bourdieu, and Graff," in *Reorientations*, ed. Bruce Henricksen and Thais Morgan (Urbana: University of Illinois Press, 1990), pp. 248–67; Robert Con Davis, "Freud, Lacan, and the Subject of Cultural Studies," *College Literature* 18, 2 (1991), 22–37; and Nancy Mergler and Ronald Schleifer, "Cognition and Narration: Binary Structures, Semiotics, and Cognitive Science," *New Orleans Review* 17, 1 (1991), 64–75.

RONALD SCHLEIFER
ROBERT CON DAVIS
NANCY MERGLER

Norman, Oklahoma

Culture and Cognition

INTRODUCTION

Science, Cognition, and Culture

Cognition and Semiotics

In this book we are attempting to bring together conceptions of cognition as they have been developed, independently, in the cognitive sciences and in semiotics in the twentieth century. The tradition of the cognitive sciences—an empirical Anglo-American tradition—has developed recently in many disciplines ranging from computer science to experimental psychology. It is, as Howard Gardner notes in *The Mind's New Science*, "a contemporary, empirically based effort to answer long-standing epistemological questions—particularly those concerned with the nature of knowledge, its components, its sources, its development, and its deployment" (1985: 6). In this effort, cognitive science raises questions about "mental" phenomena that were rarely considered in scientific and empirical psychology in the early years of the twentieth century, which was dominated by logical positivism and behaviorism. Similar long-standing questions concerning the nature and functioning of knowledge are addressed by the Continental tradition of semiotics that developed first in Prague and then in Paris. This tradition grew out of the revolutionary reconception of linguistic science arising throughout Europe (in Geneva, Moscow, and Copenhagen, as well as Paris and Prague) in the first third of the century. Unlike cognitive science, the tradition of Continental semiotics pursues a rationalist rather than an empiricist program. Beginning with language—and the intrinsic intelligibility of language—rather than with behavior, Continental semiotics assumes that knowledge and understanding can be understood and accounted for through an understanding of the logic of signification and discourse. Cognitive science, on the other hand, aims to understand behavior—cognitive activity—within an economy of other measurable behaviors. Of course, these programs change and alternate: semiotics attempts to account logically for the seemingly empirical referentiality of understanding within its science; and cognitive science follows the reason and logic of "mind" in its survey of seemingly external "data."

The purviews of these two disciplines are so disparate that people working in one are rarely aware of the work and vocabularies of the other. Yet our purpose in articulating the assumptions and methods of the two is more than simply pedagogical. We hope to develop what Bruno Latour and others studying the sociology of scientific knowledge call a "superimposition" of different descriptions of ways of knowing in order to articulate what he also calls a "network" of cognitive activities that "underwrite" a particular way of understanding cognition. That is, the aim of *Culture and Cognition* is to demonstrate the ways that the purported logic of the Continental conception of cognition and seemingly objective data gathered to support Anglo-American descriptions of the functioning of cognition can both contribute to a wider understanding of cognitive activity. That understanding of cognition—including the seemingly immediate apprehensions of knowing—situates it as an *instituted* activity that always takes place within a network of cultural assumptions, a cultural horizon of the possibilities of apprehension altogether. Such a network, we believe, is the site of the meeting of mind and world, a kind of logical empiricism—or what Richard Rorty calls "epistemological behaviorism" (1979: 174)—in which neither the reasons of mind nor the forces of the world are fully distinct from their opposites.

In this aim, a chief, if often implicit, focus of the book is on what seem to be *self-evident truths,* including the conditions of the appearance of such self-evidence. Both Continental semiotics, focusing as it does on the phenomenal "evidence" of *meaning* determined by the logical activities of "mind" (the way evident meaning is articulated) and Anglo-American cognitive science, focusing on freestanding empirical *truths* (the way truth stands "outside" apprehension) suggest that the self-evidence of cognition is in one way or another simply "given." Recent philosophy in the Anglo-American tradition—Wilfrid Sellars's critique of the "Myth of the Given" (1963: 127–96), Willard van Orman Quine's critique of the "two dogmas of empiricism" (1961: 20–46), Donald Davidson's elaboration of the "third dogma" of empiricism (1974: 11), Thomas Kuhn's work on the history and philosophy of science (1970, 1977), Rorty's critique of the epistemological tradition in philosophy (1979)—has critically examined the conception of the "givenness" of empirical data. At the same time, recent work in the Continental poststructuralist tradition by Jacques Derrida (1976, 1978), Michel Foucault (1972a, 1972b), Jacques Lacan (1977b), and a host of others has critically examined the "givenness" of phenomenal experience. Our intention is to pursue this critique of "self-evidence" in relation to Anglo-American and Continental examinations of cognition within what Davidson describes as two main conceptual schemes: the self-evidence of "objects" in the world of "reality (the universe, the world, nature)" and the self-evidence of "experience (the passing show, surface irritations, sensory promptings, sense data, the given)" (1974: 14). In the Continental tradition Paul Ricoeur, examining the nature of time, traces this opposition from Aristotle and Augustine as the opposition between cosmological time and phenomenological time (or "cosmic time" and "lived time" [1988: 245, 99]). Our intention, then, is to explore the nature of "self-evidence"—the foundations of these two traditions studying cognition—as an object of inquiry rather than its ground and starting place.

To this end, throughout *Culture and Cognition* we use what A. J. Greimas has developed (in the Continental tradition of semiotics) as the "semiotic square," a representation of the logical entailments, the "network," of the "given." This square attempts to map with logical rigor the elements that constitute the cognitive understandings of meaning. Drawing on a tradition of rational critique of the cultural sign, Greimas developed the logical basis of the square in *Structural Semantics* (first published in 1966), specifically in the penultimate chapter in which he attempts to account for the understanding of narrative discourse. Narrative discourse is another major focus of our book. A Greimassian description of the cognitive aspects of narrative is the focus of Chapter 2, and, as we mention in Chapter 3, where we describe cognitive understanding in terms of an empirical examination of the ways in which the species adapts aging to cultural-cognitive ends within the ecology of human life, narrative discourse makes experience in time meaningful. In our reading of Greimas, the concept of narrative potentially tempers the rigors of Continental rationalism in the same way that Greimas's semiotic logic tempers the commonsensical assumptions of empiricism. The importance of Greimas's rigorous analysis set forth in Chapter 2 is its attempt to account for narrative meaning in terms of cognitive structures. The complementary aim of Chapter 3 is to account for empirically measured cognitive activity—specifically, to argue for the adaptiveness of the discursive formations of aging—in terms of the explanatory narrative of what Stephen

Jay Gould calls "cultural evolution" (1981: 324). In both cases, these different strategies create "an authorizing center of meaning, precisely, a narrative shaping of natural history" that Eric White describes (1990: 101).

The "strategy" of Greimas's semiotic square, however, grows out of a philosophical tradition that encompasses both Continental rationalism and Anglo-American empiricism. This tradition, which can be traced from the pre-Socratic philosophers up through contemporary thinkers, is not particularly concerned with narrative discourse. Rather, it assumes that the structures and mechanisms of cognition—of understanding altogether—simply and immediately apprehend logical relationships that exist empirically in the world. A basic assumption of empirical science is implicit in this Western tradition, namely that matters of fact are governed by the same reason that mind uses and apprehends. It is precisely here that "common sense" takes its stand: what is "common" about common sense is its universality, transcending every particular occasion in an essential human nature "fitted" to the world. (For a discussion of Kant's conception of common sense in similar terms, see Deleuze 1984: 21–27; for a discussion of Kant's relationship to science, see Rorty 1982: 92.) Thus, in *The Ideology of the Aesthetic* Terry Eagleton notes that "the harmony of faculties [in Kant] which is aesthetic pleasure is in fact a harmony requisite for every empirical cognition; so that if the aesthetic is in a sense 'supplementary' to our other activities of mind, it is a supplement which turns out by some Derridean logic to be more like their foundation or precondition" (1990: 102). In other words, empirical cognitive science assumes that cognition is the immediate apprehension of logical relationships, which are "thought"—as the mind reproduces transcendental categories of empirically existing relationships—in terms of logical binary oppositions that account for meaning and cognition by simply recognizing abstract logical relationships within or across concrete data. Alfred North Whitehead sums up this tradition in his discussion of the nature of mathematics in *Science and the Modern World* when he says that mathematics "is a resolute attempt to go the whole way in the direction of complete analysis, so as to separate the elements of mere matter of fact from the purely abstract conditions which they exemplify" (1967: 24). Such analysis, he writes, "enlightens every act of the functioning of the human mind," and it does so in terms of "the direct aesthetic appreciation of the content of experience," in terms of the apprehension of "the absolutely general conditions" governing the relations of the elements of the content of experience, and finally in terms of the "variety of occasions" of experience itself (1967: 24–25).

Whitehead is describing three conditions governing (or characteristics describing) "scientific" understanding: the attempt science makes to present an understanding of phenomena that is simple, exhaustive, and generalizing. We are arguing that Continental semiotics, growing out of a rationalist tradition, emphasizes the *simplicity* of understanding—its parsimonious logical coherence—whereas Anglo-American empiricism emphasizes the attempt of understanding to account for matters of fact as *exhaustively* as possible. The *generalizing* aspect of understanding presents a more complex criterion. Its generalizations suggest the predictive power of understanding, the ability of understanding to make sense of facts not yet encountered. (This will be especially important to our discussion of the emergence of knowledge in Chapter 7.) However, the generalizations of understanding also include the social fact that understanding will be generally accepted, and so they encompass the relationship of culture to cognition. Rorty examines this relationship by discussing what Kuhn describes as the generalizing criterion of "scope" that governs scientific understanding in the most global way by marking "the lines between disciplines, subject matters, [and] parts of culture" (Rorty 1979: 329; Kuhn 1977: 322). In Chapter 2 we present an example of cognitive simplification in a Greimassian analysis of the nature of the genres of literary narratives. In Chapter 3 we present an example of an empirical survey of experimental data that attempts to test, as exhaustively as possible, the cognitive functioning of old people. In Chapters 5 and 6 we return to narrative understandings of psychology and aging from the vantage of social generalization—specifically, to the psychological subject in a Freudian case history, to the narrative subject in Einstein's explanation of relativity, and to the rhetoric of old people in interviews and in poetry. In each case we are examining the relationship between culture and cognition, discourse and understanding.

In his description of the nature of mathematics, Whitehead argues that mathematics most fully satisfies the three aspects of cognition in that it apprehends experience "in the manner of a pattern with a key to it" (1967: 26). That key, he suggests, is "the harmony of the logical reason, which divines the complete pattern." This "reasonable harmony" is the harmony of mind and world so that "thought can penetrate into every occasion of fact [and] by comprehending its key conditions, the whole complex of its pattern of conditions lies open before it" (1967: 26). The key to this understanding is that cognition apprehends what is universally and transcendentally true, what in no way is determined or affected by the act of cognition itself. It is possible that the act of cognition can struggle for its results, pursue false paths, make mistakes in assumptions and calculations

(1967: 22). But what is important and remarkable about cognition for Whitehead—what literally *constitutes* understanding—is that true understanding, finally, is disinterested. Only one thing can be said of the possibility of the irrationality of existing things, Whitehead asserts, and that is that nothing can be said of them. When understanding is achieved, its objects are—they have to be—self-evident in their truth. "Either we know something of the remote occasion [related to the immediate occasion at hand] by the cognition which is itself an element of the immediate occasion," he writes, "or we know nothing. Accordingly the full universe, disclosed for every variety of experience, is a universe in which every detail enters into its proper relationship with the immediate occasion. The generality of mathematics," he concludes, "is the most complete generality consistent with the community of occasions which constitutes our metaphysical situation" (1967: 25). For Whitehead, mathematical "generality" is simple, not complex, since the "metaphysical situation" of the relationship between mind and world is not complex. The empirical universe itself, Whitehead is arguing, is one of reasonable, "proper" relationships in which the same laws govern all details, no matter how immediate or remote they may be. In such a universe, the activity of mind—of understanding—in no way affects the objects of cognition: mind is separate from empirical "data," which exist separately and apart. In this way, cognition is a noncomplex activity that literally has no *interest* in what it studies.

Against such an abstract description of knowledge—one in which the "objects" of the world are simply "given" to cognition—Greimas's semiotic square attempts to map the nature of logical relationships that *conditions* particular cognitive (i.e., semantic) understanding. The relationships found in the square—relationships of contrariety, contradiction, and implication, which exhaust the possibilities of "relationship"–describe the logic that Whitehead mentions but does not analyze. As in Whitehead, these relationships, finally, are without interest: their existence does not participate in the phenomena they govern. But whereas Whitehead assumes that the laws of logic are general in *all* matters of fact, Greimas assumes that logic simply governs *all* cognitive apprehension of meaning. Such apprehension, as Greimas himself notes, distinguishes between logic and semantics (1970: 12). As Quine says in a very different context, "things had essences, for Aristotle, but only linguistic forms have meanings. Meaning is what essence becomes when it is divorced from the object of reference and wedded to the word" (1961: 21; see also Gardner 1985: 361–70 for empirical studies that suggest this distinction).

Behind Greimas's articulation of the logical relationships of the semiotic square—they are all binary relationships—is the simplifying assumption, borrowed from Continental phenomenology, that cognition and meaning are felt to be "immediate" givens of experience. As such, it is *phenomena*—that is, *appearances to mind*—that are the stuff of cognition. For this reason the description of "the conditions [of] the reception of signification" in discourse is of the utmost importance for Greimas. "Although the [discursive] message," he writes, "is presented for reception as an articulated succession of significations, that is to say, with diachronic status, the reception can be effectuated only by transforming the succession into simultaneity and the pseudo-diachrony into synchrony. Synchronic perception, if one believes Bröndel, can apprehend only a maximum of six terms at the same time" (1983b: 144). For Whitehead, every "occasion of fact" in the universe participates in meaningful relationships susceptible to cognition. For Greimas, the "reception" and "perception" of signification or meaning—the very recognition of "meaningful relationships"—is governed by the mind's apprehension of simultaneous, logical structures. Here, as elsewhere in Greimas, "reception" and "perception" describe cognition and cognitive activity.

In this account we can see that Davidson's opposition between "objects" and "experience" that we are following in contrasting the Continental focus on meaningful experience to Anglo-American empiricism is not absolute. For instance, Greimas's citation of Viggo Bröndel's assertion of the "maximum" number of terms of relationship that can be apprehended simultaneously finds its counterpart in the important review of empirical studies by George Miller, "The Magical Number Seven, Plus or Minus Two: Some Limits on Our Capacity for Processing Information," that Howard Gardner describes as having "a decidedly major impact" on the emerging cognitive sciences (Miller 1956; Gardner 1985: 90). The two descriptions of cognition we have presented in Whitehead and Greimas delimit a common ground in their descriptions of how cognition functions, even if they arrive at it through the divergent paths of empirical study and logical implication.

Still, that divergence is clearly marked in the self-conscious "narrative" element of the semiotic tradition, the focus on discourse and signs as constitutive in cognition rather than its secondary epiphenomenon. This is particularly clear in a narrative sketch of the relationships Greimas inscribes in the semiotic square. Whenever we have a concept (or any semantic content), Greimas suggests, we understand it in relation to its opposite. Thus, for example, the concept of /reason/ suggests its contrary, /irrationality/. But this opposition exists on a particular "axis" of understanding, because other concepts besides /irrationality/ could be opposed to /reason/ (e.g., /feeling/). Thus, in a full comprehension of this concept, its contradictory, is also suggested. A contradictory relationship offers the absence to the presence of the defining quality of the initial concept. For instance, if /irrationality/ is the contrary to /reason/, then its contradictory—the *absence* (as opposed to the negation) of reason—might be called /intuition/, immediate comprehension as opposed to reasoned comprehension. This third concept, insofar as it presents the contradictory to the first, also describes the axis on which the contrary relationship, in this example reason vs. irrationality, inscribes itself. That is, the semantics of /intuition/ suggests the immediacy of irrationality and, at the other extreme, the mediacy of reason. (In a contrary relationship, the opposing elements can be understood as the extreme ends of an axis or continuum.) The third term describes this axis by sharing some quality with its contradictory (hence, the fact of *relationship* altogether). In this instance, this quality is that of /mediation/ or /articulation/: intuition is the absence, and, *in an important way,* reason is the presence of the mediation of articulation. By marking the absence of mediation, /intuition/ (as an element on the square) names the axis or continuum of which reason vs. irrationality are the extreme ends: the axis of the givenness of apprehension (ranging from the precise and consistent mediated articulations of reason to the imprecise and inconsistent immediacy of unreasonable flux). Finally, the fourth element of a semiotic square is the contrary to the third: in this case, it is the mediated articulations that are the contrary to immediacy of /intuition/, what we will call /discourse/. Here is the square we have described:

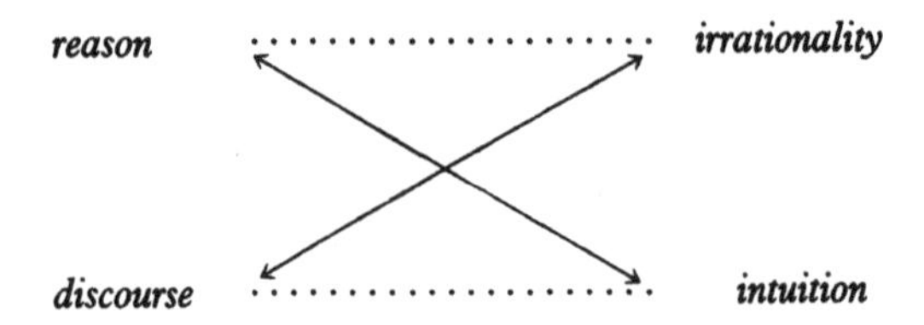

When we said the third element defines the first "in an important way," we meant that by suggesting a concept that encompasses (semantically) the axis on which the first element defines itself against its contrary, it presents a new framework in which to understand the initial concept.

The fourth corner of the square makes that new framework explicit. In this instance, it suggests that the abstract qualities of logic and precision that inhabit the concept of /reason/ in a self-evident way—we have already seen that self-evidence in Whitehead's description of the metaphysics of mathematical understanding—are themselves contingent upon (i.e., both constituted and instituted by) a network of understanding. Thus, the semiotic square "unpacks" self-evident concepts, and in its very workings it suggests that self-evident truths always exist within a network of understandings so that cognition—even the self-evidence of the *immediate* cognitive forms of reason and unreason—is never simply given, never simply "data." For instance, a native speaker of English immediately recognizes the difference (if not opposition) between the phoneme /l/ and the phoneme /r/: we immediately *perceive* the difference between "light" and "right." Native Chinese speakers, however, have great difficulty with this "perception": in most Chinese dialects, the sounds [l] and [r] are variations on the "same" phoneme, what linguists call allophones. In English an aspirated and unaspirated [t] are allophones of /t/, as are [t] sounds pronounced at different pitches. In Vietnamese, to take one instance, such differences of pitch distinguish between phonemes and are *perceived* as clearly different as our /l/'s and /r/'s. In these phonological examples, we see the cultural institution of so-called immediate perceptive cognition. In the same way, the logic of the semiotic square allows us to see the instituted nature of immediate cognitive apprehensions. It allows us to see that even self-evident truths (e.g., that "reason" is simply precise logic) exist within a framework of complications and assumptions—in this instance, the unperceived complication between reason and linguistic and discursive activities of articulation.

This logical "unpacking," however, presents dangers to an understanding of the activity of cognition equal to that of the more or less mentalist metaphysics Whitehead presents. It does so by privileging the linguistic sign as foundational for cognition in the same way Whitehead privileges the "objective" truth and the harmony of thought with that truth. Whereas Whitehead makes "thought" or "mind" *simply* a passive or harmonious response to the world, Greimas builds his understanding of cognition—an understanding which is representative of semiotic understandings in general—on the basis of the nature and functioning of linguistic signs that are complicated but finally not complex. For Greimas, signs are understood in terms of the same logical binary oppositions that govern models of cognition in the Anglo-American tradition of cognitive science. That is, even though Greimas (as well as Saussure and Lévi-Strauss, whom he follows in his semiotic analysis) makes note of the social constitution of signs, the social-cultural activity of cognition is submerged in his logico-semantic analyses. In fact, one benefit of what Bruno Latour calls the "superimposition" of the Anglo-American assumption of "mind" and the

Continental assumption of the self-evidence of "signs" is the discernment of the force of culture in activities of cognition.

The concept of "superimposition" helps define the functioning of narrative understanding within discourse. In his remarkable study *Time and Narrative,* Paul Ricoeur analyzes the mimetic and cognitive functioning of narrative in terms of the "ideas of the Same, the Other, and the Analogous" (1988: 143). These "ideas," in important ways, are delineated within the semiotic square and allow us to describe the square (as we have) as a "narrative sketch." In our narrative, we chose *reason* to begin with—a semantic value seemingly simple and "self-same"—and generated its "Other," *irrationality.* Then we posited a broader category of *intuition* of which *reason* and *irrationality* became "parts"—in this way species of the "Same." Finally, we named *discourse* as the fourth corner of the square, a concept which is "Analogous" to *reason,* positing similarity that is, finally, an act of judgment. Narrative, Ricoeur argues, accomplishes such judgments—he calls them the "synthesis of the heterogeneous" (1984: ix)—by organizing a series of events "into an intelligible whole, of a sort such that we can always ask what is the 'thought' of this story" (1984: 65). Narrative seeks neither the simplicity of the Same, Whitehead's "abstract conditions" exemplified by the "data" of the world, nor the multiplications of Otherness, organized, a priori, by semiotic's generative principle of difference. Instead, it "understands" by means of configurations of parts and whole. If the Anglo-American tradition aims at empirically revealing the Same within data, and if Continental semiotics emphasizes the constituting power of difference and Otherness, narrative configures disparate elements by means of "the Analogous, which is a resemblance between relations rather than between terms per se" (1988: 151).

In describing narrative, Ricoeur examines all three of these modes of cognition, what he calls "theoretical," "categoreal," and "configurational" modes of comprehension. "According to the theoretical mode," he writes,

> objects are comprehended in terms of a case or as examples of a general theory. The ideal type of this mode is represented by Laplace's system. According to the categoreal mode, often confused with the preceding one, to comprehend an object is to determine what type of object we are dealing with, what system of a priori concepts organizes an experience that otherwise would remain chaotic. Plato aims at this categoreal comprehension, as do most systematic philosophers. The configurational mode puts its elements into a single, concrete complex of relations. It is the type of comprehension that characterizes the narrative operation. All three modes do have a common aim, which is no less implicit in the configurational mode than in the other two. Comprehension in the broad sense is defined as the act "of grasping

> together in a single mental act things which are not experienced together, or even capable of being so experienced, because they are separated by time, space, or logical kind. And the ability to do this is a necessary (although not a sufficient) condition of *understanding*." (1984: 159; Ricoeur is citing Mink 1970: 547)

For Ricoeur, narrative configurations—the apprehension of temporal wholes and elemental "events" from serial and disparate phenomena or data—accomplish the "superimposition" that Latour describes and that we pursue in this book. Such superimpositions, like narrative itself, do not reduce difference to the same, nor do they satisfy themselves with the contingency and accident of arbitrary likenesses. Narrative "unifies into one whole and complete action the miscellany" of "circumstances, ends and means, initiatives and interactions, the reversals of fortune, and all the unintended consequences" of action (1984: x) without condemning the elements of that miscellany—the congeries of "circumstances" and "events" enumerated—to being lost in its "unity."

Objectivity

To the extent that such superimposition is possible, the analysis focusing on the "experience" of cognition—the very *meanings* of "comprehension" and cognitive understanding captured in the semantics of "intuition"—can also be extended to cognition understood not as "articulating" self-evident experience, but as "accounting for" self-evident objects in the world. In such an extension, *meaning* would be replaced by *truth* so that cognition would be measured against its object rather than its subject. The opposition between objectivity and subjectivity is an important—and recurrent—one. Whereas "meaning"—especially species of Kant's "analytic" meaning which seem to inhabit the binary oppositions of the first two (or three) corners of Greimas's square—is closely related to the logical *simplicity* of scientific knowledge, "truth" or "reference" is closely related to the empirical project of accounting for as much of the world—as much "data"—as possible.

Such an extension can be seen in Thomas Kuhn's defense of his influential thesis describing the structure of scientific revolutions. That structure, he argues, consists of a "paradigm" of "normal science"—a realm of shared assumptions about the criteria for recognizing valid scientific explanations, which Kuhn later describes as that which "members of a scientific community . . . share," a "disciplinary matrix" allowing the community to function "as a producer and validator of sound knowledge" (1977: 294, 297, 298). In the history of science scientific revolutions occur when such paradigms are replaced by new matrices of understanding. "In a sense I am unable to explicate further," Kuhn writes in *The Structure of Scientific Revolutions*,

> the proponents of competing paradigms practice their trades in different worlds. One contains constrained bodies that fall slowly, the other pendulums that repeat their motions again and again. In one solutions are compounds, in the other mixtures. One is embedded in a flat, the other in a curved matrix of space. Practicing in different worlds, the two groups of scientists see different things when they look from the same point in the same direction. Again, that is not to say that they can see anything they please. Both are looking at the world, and what they look at has not changed. But in some areas they see different things, and they see them in different relations one to the other. That is why a law that cannot even be demonstrated to one group of scientists may occasionally seem intuitively obvious to another. Equally, it is why, before they can hope to communicate fully, one group or the other must experience the conversion that we have been calling a paradigm shift. . . . Like the gestalt switch, it must occur all at once (though not necessarily in an instant) or not at all. (1970: 150).

Richard Rorty takes exception to Kuhn's description here of "different worlds." Such a description, he argues, skirts with a "fall" into idealism suggested by the opposition between the mind's "making" vs. the mind's "finding" nature (1979: 244), and it does so by participating in what he describes in *Philosophy and the Mirror of Nature* as the tradition of epistemology in philosophy, "the Cartesian . . . triumph of the quest for certainty over the quest for wisdom" (1979: 61).

This tradition is based on—it creates—the opposition between idealism and empiricism. It creates the alternatives of conceiving of knowledge as either a function of mind or a function of the world. For this reason, Rorty cites C. I. Lewis's description of "one of the oldest and most universal philosophical insights," namely that "there are, in our cognitive experience, two elements; the immediate data, such as those of sense, which are presented or given to the mind, and a form, construction, or interpretation, which represents the activity of thought" (cited by Rorty 1979: 149). Rorty goes on to argue that this "insight" is neither old nor universal, but rather inhabits the tradition of epistemological thinking that assumes that "cognitive experience" is self-evident (see 1979: 149–50). That is, the tradition Rorty analyzes is founded, as Richard Bernstein describes it, on "the idea of a basic dichotomy between the subjective and the objective; the conception of knowledge as being a correct representation of what is objective; the conviction that human reason can completely free itself of bias, prejudice, and tradition; the ideal of a universal method by which we can first secure firm foundations of knowledge and then build the edifice of a universal science; the belief that by the power of self-reflection we can transcend our historical context and horizon and know things as they really are" (1983: 36). In short, this is the tradition Whitehead assumes in his description of scientific knowledge as a harmony of mind and world. It is also a tradition that, in its assumption of simple "objectivity," governed the work of much experimental psychology in the twentieth century, such as that of James J. Gibson who argued that perception is simple and immediate. Implicit in this argument, as Gardner describes it, is the "belief in the real world as it is, with all the information there, and the organism simply attuned to it" (1985: 317; see 308–18).

Most of the critics who attacked Kuhn's thesis were less sophisticated and subtle than Rorty is in his critique and leveled the charge of "subjectivity" against him. (See Bernstein 1983: 22–25 for a survey of the "storm of protest" Kuhn's thesis occasioned. Kuhn 1977: 321 also surveys the protest.) Such a charge, Kuhn argues in his defense, is confused. "'Subjective,'" Kuhn argues, "is a term with several established uses: in one of these it is opposed to 'objective,' in another to 'judgmental.' When my critics describe the idiosyncratic features to which I appeal as subjective, they resort, erroneously I think, to the second of these senses. When they complain that I deprive science of objectivity, they conflate the second sense of subjective with the first" (1977: 336). Kuhn goes on to describe the second sense of "subjective" as being like "sensation reports," reports which are "matters of taste [and so] are undiscussible." Opposed to this sense of "subjective" are "judgments" which, in fact, can be discussed precisely in terms of the bases for judgment: if I report I did not enjoy a movie, there is no discussion; if I judge that it was "a pot boiler," my judgment can be argued and discussed (1977: 336–37). The first sense of "subjective" is significantly different. "Whether my taste is low or refined," Kuhn writes, "my report that I liked the film is objective unless I have lied. To my judgment that the film was a pot boiler, however, the objective-subjective distinction does not apply at all, at least not obviously and directly. When my critics say I deprive theory choice of objectivity, they must, therefore, have recourse to some very different sense of subjective, presumably the one in which bias and personal likes or dislikes function instead of, or in the face of, the actual facts" (1977: 337).

Kuhn is arguing, however, that a definition of "objectivity" as actual "facts . . . independent of theory" is itself a problem that must be examined. "Proponents of different theories," he argues,

> are . . . like native speakers of different languages. Communication between them goes on by translation, and it raises all translation's familiar difficulties. The analogy is, of course, incomplete, for the vocabulary of the two theories may be identical. . . . [S]ome words in the basic as well as in the theoretical vocabularies of the two theories—words like "star" and "planet," "mixture" and "compound," or "force" and "matter"—do function differently. Those differences are unexpected and will be discovered and localized, if at all, only by repeated experience of communication breakdown. (1977: 338)

Other terms beyond those of physics that Kuhn chooses, terms like "cognition," "perception," "truth," and "meaning," but also what Gardner calls the "most prototypical cognitive relations—like believing, expecting, thinking, and so on" (1985: 313)—will also present global and local differences. One such term is Kuhn's word "subjective," and it is the contention of Continental semiotics that the complexities of the meanings of such a term are not quite as "unexpected"—not quite as simple an *empirical* accident—as Kuhn in his notion of "shift" or Rorty in his notion of the "unpredictability" of "abnormal discourse" (1979: 320) suggests. As we suggest in the discussions of Freud later in this book, "communication breakdown" is not simply accidental, but the very motor of psychoanalytic cognitive understanding.

In this example, however, by distinguishing "subjective" from both "objective" and "judgmental," Kuhn is "exploding," as Greimas says, the semantics of /subjective/ in precisely the way that the semiotic square maps meaning.

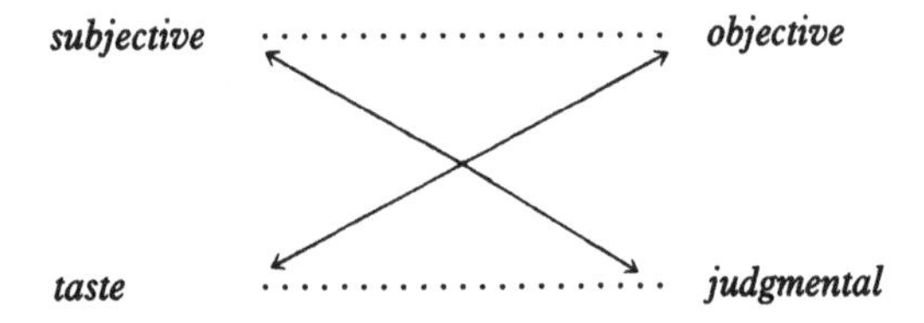

This square describes Kuhn's first and second senses of "subjective": subjective as opposed to objective and subjective as opposed to judgmental. Moreover, insofar as "taste" presupposes subjectivity and "judgment" presupposes objectivity, it also maps the relation of presupposition between the levels Greimas also describes in the semiotic square. Finally, in the opposition between "taste" and "objectivity" the square describes the opposition Davidson described between "experience" and "reality." "Taste," as Kuhn describes it, is simply and phenomenally given, the contrary to judgment and the contradictory to empirical "data" of objectivity.

What Kuhn doesn't recognize is that by emphasizing the *phenomenological* feature of the complex semanticism of /taste/, he could have opposed it as the contrary to the *empirical* feature of /objective/ on the level (or "axis") of the criteria of scientific understanding. With this emphasis, Kuhn would have marked the two dogmas of empiricism that Quine describes, the dogma of the self-evidence of the analytic-synthetic opposition that corresponds to the phenomenology of *meaning,* and the dogma of the self-evident reduction of significance to immediate *empirical* experience (Quine 1961: 20). Such an emphasis would have unpacked the "standard criteria for evaluating the adequacy of a [scientific] theory" Kuhn describes in his own defense, namely the more or less empirical list, "accuracy, consistency, scope, simplicity, and fruitfulness" (1977: 322). This list is reducible to the three criteria that describe scientific knowledge we mentioned already, simplicity (which includes self-consistency), exhaustiveness (which includes "accuracy"), and generalization (which includes "scope," "fruitfulness," and the conception of "consistency" that suggests understanding must be consistent with the understanding of other objects of understanding). If simplicity and exhaustiveness correspond to the two dogmas of empiricism Quine describes, generalization corresponds to the "third dogma" of empiricism that Davidson articulates, the opposition between "scheme and content, of organizing system and something waiting to be organized" (1974: 11), the very conception of "paradigm" that governs Kuhn's argument.

This third dogma offers a final example of the functioning of Greimas's square. If we begin with the self-evidence of matters of fact and map the three elements of scientific understanding that we have mentioned, we will get the following square:

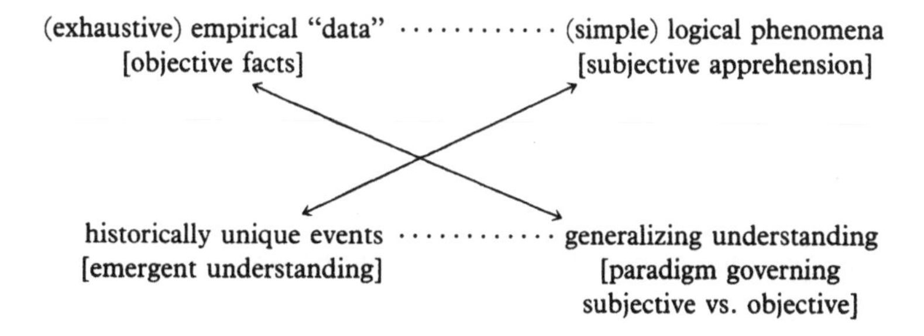

In this square the *complexity* of the third term becomes clear. Earlier we marked that complexity in the combination of the immediacy and inarticulateness of /intuition/: "intuition" is the opposite of demonstration, and more or less figuratively, it suggests inarticulate apprehension, a figurative inability to articulate the very "ideas" that are intuited. In a similar way, but without what some might call the sleight of hand of switching from literal to figurative meaning, /generalizing understanding/ exhaustively accounts for data—as Whitehead says, generalization assumes a universe in which "every detail enters into its proper relationship" with the whole of the universe—and, at the same time, it simplifies the experience of such a universe of details. Here we can see that the third term of the Greimassian square is always complex: as Greimas himself says, it explodes the unity of self-evidence (1988: 45). This analysis also accounts for the double sense of "consistency" we outlined earlier: generalization combines rational and empirical "consistency."

The fourth term of the square is also complex, but negatively so. It is neither an empirical "detail" nor a "phenomenal" experience but rather an "event" that entails the intersection—the superimposition—of objective fact and phenomenon. The concept of "event," in Ricoeur's terms, is configurational: "just as it is possible to compose several plots on the subject of the same incidents (which, thus, should not really be called the same events), so it is always possible to weave different, even opposed, plots about our lives" (1988: 248). As we argue in Chapter 4, the fourth term allows us to reimagine the cognitive activity of *referring* to the world in both the assumption of "objectivity" in empiricism and the occlusion of reference by meaning in semiotics. The fourth term of this square thus reorients our "self-evident" understandings of "data" and "phenomena" by erasing the opposition between mind and world in the concept of events in which neither—in the simplicity of "mind" or "world"—participate. In this framework, even cognition is an event, that is, "cognitive activity." The empirical tradition emphasizes the activity of cognition—after all, it studies and measures "behavior"—even while it rarely questions the effects of the behavior of the scientific "observer" on his or her data. The semiotic tradition, on the other hand, emphasizes the participation of its practitioners in semiotic activity even while it assumes (with phenomenology) the ineluctable "givenness" of meaning. Still, this square (unlike the square of "reason") builds its understanding of cognition—an understanding representative of empirical understandings in general—on the basis of preexisting "facts."

Thus, the analysis of scientific understanding mapped by this square suggests a critique of what Rorty, following Sellars, Quine, and Davidson, calls the "myth of the given." In Rorty, as in Sellars, the "given" entails both the phenomenally given and the empirically given—the myth of the absolute ineluctability of sense-data and/or their objects. What remains "given" in Rorty and Sellars, however, is the ineluctable "givenness" of historical event, what Rorty calls the ongoing "conversation" of philosophy. Even Sellars, in his critique of the myth of the given, makes the temporality of understanding important. "If I reject the framework of traditional empiricism," he writes,

> it is not because I want to say that empirical knowledge has *no* foundation. For to put it this way is to suggest that it is really 'empirical knowledge so-called', and to put it in a box with rumors and hoaxes. . . .
>
> *Above all,* the picture is misleading because of its static character. One seems forced to choose between the picture of an elephant which rests on a tortoise (What supports the tortoise?) and the picture of a great Hegelian serpent of knowledge with its tail in its mouth (Where does it begin?). Neither will do. For empirical knowledge, like its sophisticated extension, science, is rational, not because it has a *foundation* but because it is a self-correcting enterprise which can put *any* claim in jeopardy, though not *all* at once. (1963: 170)

By alluding to William James's tortoise and Hegel's serpent, Sellars is situating both the Anglo-American tradition of empirical *fact* and the Continental tradition of rational *meaning* within the temporal order of cultural activity.

Such an order, as Rorty's figure of "conversation" suggests, is "given" in a very different manner than the "givenness" of empirical data and experiential phenomena. It is given in the form of what Quine calls "cultural posits" (1961: 44). Quine narrates the history of philosophy—it is really Anglo-American philosophy—by describing the reorientation in semantics from Locke and Hume to Bentham, Frege, and Russell "whereby the primary vehicle of meaning came to be seen no longer in the term but in the statement" (1961: 39). This latter idea of "defining a symbol in use was, as remarked, an advance over the impossible term-by-term empiricism of Locke and Hume. The statement, rather than the term, came with Bentham to be recognized as the unit accountable to an empiricist critique. But what I am now urging is that even in taking the statement as unit we have drawn our grid too finely. The unit of empirical significance is the whole of science" (1961: 42). The whole of science Quine describes is the historically unique "paradigm" within which understanding is possible and cognition takes place. But these two locutions of "understanding" and "cognition" are not necessarily or fully commensurate. Rather, they articulate, in Quine's words, "two competing conceptual schemes, a phenomenalistic one and a physicalistic one" (1961: 17)—the very "schemes" we have been discussing as the Continental and Anglo-American traditions (two of Ricoeur's three categories). Quine goes on to suggest that both schemes have advantages: "the physical conceptual scheme simplifies our account of experience because of the way myriad scattered sense events come to be associated with single so-called objects"; while "from a phenomenalistic point of view, the conceptual scheme of physical objects is a convenient myth, simpler than the literal truth and yet containing the literal truth as a scattered part" (1961: 17, 18).

The point is that Quine's "cultural posit"—like the /historically unique events/ we are examining—is not simple or exhaustive or general in any usual sense of the word. Rather, like "tradition" itself, it takes the form of narrative understanding and narrative argument. "A tradition," Alisdair MacIntyre notes, "not only embodies the narrative of an argument, but is only to be recovered by an argumentative retelling of that narrative which will itself be in conflict with other argumentative retellings. Every tradition therefore is always in danger of lapsing into incoherence and when a tradition does so lapse it sometimes can only be recovered by a revolutionary reconstitution. Precisely such a reconstitution of a tradition which had lapsed into incoherence was the work of Galileo" (1977: 461; see also Bernstein's citation of MacIntyre's discussion of case-histories in the unpublished version of this article [1983: 57]). Such narrative is not—or not simply—a "constructed" world like that which Kuhn seems to imply; it is not simply a "subjective" relativist accounting of understanding or cognition. But neither is it simply an accounting of what is.

That is, the narrative of understanding cannot be reduced to either the construction of mind or the recording of what is. Rather, it is irreducibly complex. Citing Norbert Wiener's description of information theory, Gardner articulates this complexity: "Information is information, not matter or energy," Wiener wrote in *Cybernetics, or Control and Communication in the Animal and the Machine;* "no materialism which does not admit this can survive at the present day" (cited in Gardner 1985: 21). In the terms of Shoshana Felman, cognitive knowledge—what she calls the "referential knowledge of language"—"is not knowledge *about* reality (about a separate and distinct entity), but knowledge that *has to do with reality,* that acts within reality" (1983: 77). Such "activity," like all *acts*—and like the *narrativity* of discourse and understanding that calls for a "semantic syntax" rather than a "logical syntax" (see Greimas 1983b: 149–50)—can be apprehended both globally and atomistically. As Davidson suggests, the concept of "act" is irreducibly complex precisely because "events" can be analyzed globally (e.g., the queen killed the king) and atomistically (e.g., the queen moved her hand, poured the poison, and the poison killed the king a week later by affecting his nervous system in such and such a way [which itself can be analyzed in greater and greater detail]) (see Davidson 1980: 57–61; and Ricoeur 1984: 122–23). The very apprehension of "events," like the apprehension of phonemes, configures simple unities from a larger number of elements. "The way of judging about particular cases," Ricoeur notes, "does not consist in placing a case under a law but in gathering together scattered factors and weighing their respective importance in producing the final result" (1984: 125).

Moreover, the forms of narrative comprehension have "to do with reality" atomistically and globally, and this opposition *corresponds* to that of empiricism and rationalism, but without the imperative of choosing one or the other (either knowledge "accounting for" reality or understanding "articulating" experience). That is, narrative comprehension, like narrative itself, suspends the law of the excluded middle. Instead, it offers versions of cognition—configurations, analogies, wholes that do not erase parts—that can, in fact, be "superimposed" upon one another precisely to create "middles." In Greimas's words, it is "neither pure contiguity nor a logical implication" (1983b: 244).

At the extreme of "atomistic" analysis, this conception of cognition embodies the "constrained constructivism" that Katherine Hayles describes in her narrative of the intersection between Greimassian semiotics and empirical studies of perception, what she describes as the "synergy between physical and semiotic constructs that brings language in touch with the world." "Physical constraints," she goes on, "by their consistency, allude to a reality beyond themselves that they cannot speak; semiotic constraints, by generating excess negativity, encode this allusion into language. There is a correspondence between language and our world, but it is not the mysterious harmony Einstein posited when he said that the mystery of the universe is that it is understandable. Neither is it the self-reflexivity of a world created through language and nothing but language" (1991: 83). In this account physical constraints governing understanding is a version of empiricism *negatively* conceived: "when constraints become representations, they necessarily assume a positive cognitive content that moves from" the border between the "unmediated flux" of "reality" and "the constructed concepts that for us comprise the world" into the "theater" of constructed concepts (1991: 82, 77–78). Here, then, in "constraints" one finds another language for empiricism.

At the other extreme is Rorty's version of what Richard Bernstein calls "postempiricist philosophy" (1983: 20), epistemology negatively conceived. Such philosophy is less interested in more adequately refining, as Hayles does, our sense of atomistic empirical "events" than in reorienting our conception of "understanding" globally conceived. From the time of "the triumph of the quest for certainty over the quest for wisdom," Rorty writes, "the way was open for philosophers either to attain the rigor of the mathematician or the mathematical physicist, or to explain the appearance of rigor in these fields, rather than to help people attain peace of mind. Science, rather than living, became philosophy's subject, and epistemology its center" (1979: 61). Rorty wants to return "philosophy"—and with it, if not "cognition," then at least "understanding"—to the study of living rather than science. Such a study conceives of "events" not as constraints upon understanding but as instituted by understanding. "As soon as we start thinking of 'the world,'" he says, "as atoms and the void, or sense data and awareness of them, or 'stimuli' of a certain sort brought to bear upon organs of a certain sort, ... we are now well within some particular theory about how the world is" (1982: 14). Such theories are *narrative* theories, and their alternative is *another* narrative, one Rorty borrows from Dewey. In Dewey's "historicist vision," he writes, "the arts, the sciences, the sense of right and wrong, and the institutions of society are not attempts to embody or formulate truth or goodness or beauty. They are attempts to solve problems" (1982: 16).

Samuel Weber, working more closely out of the Continental rather than the Anglo-American tradition, quotes Gaston Bachelard's version of Rorty's description of philosophy and understanding as analogous to "edifying conversation." Weber cites Bachelard's description of "the polemical character of cognition," and argues that such polemics inhabit experimentation as well as explanations so that "scientific reality" itself is "*ambivalent, agonistic,* and *conflictual*" (1987: xiii). Within the Anglo-American tradition Donald Davidson also takes a position of what we might call "unconstrained constructivism." "Having a language and knowing a good deal about the world," he writes,

> are only partially separable attainments, but interpretation can proceed because we can accept any of a number of theories of what a man means, provided we make compensating adjustments in the beliefs we attribute to him. What is clear, however, is that such theory construction must be holis-

> tic: we cannot decide how to interpret a speaker's 'There's a whale' independently of how we interpret his 'There's a mammal', and words connected with these, without end. We must interpret the whole pattern. (1980: 257)

The "whole pattern," in this examination of empirical objectivity is the issue of what Hayles means by "physical" constraint and how it interacts with the whole pattern of what we mean by understanding and/or cognition.

The conflict and polemics surrounding the given—the physical "data" of the world, the phenomenal "data" of experience—are especially pronounced in the study of cognition. In fact, it is in order to jumble or reorient received ideas that we examine both Greimas's rigorous and *analytical* analysis of narrative and the more leisurely and *narrative* analysis of cognitive events in Part I. Unlike Hayles's model of physical science, Rorty's analysis is an examination of the "urge toward unified science" inhabiting psychology in cultural-narrative terms (1979: 216–18). Similarly, Davidson examines the relations of physics to psychology in "The Material Mind" in a manner that suggests the very *languages* of these different modes of understanding, like the "translations" between different language Kuhn describes, do not necessarily "add up" to knowledge: "it is one thing," he writes, "for developments in one field to affect changes in a related field, and another thing for knowledge gained in one area to constitute knowledge of another" (1980: 247).

Superimposed Cognition

The central issue for cognitive understanding is the very notion of "adding up": for the tradition of empiricism, making sense *means* making sense of the data, adding it up in a way that accounts for the given. For the tradition of rationalism, sense making *precedes* the given: it is, in fact, what "gives" (that is, constitutes) the given. But even these notions of "analysis," inhabiting in different ways these traditions, are sites of conflict. "In spite . . . of the persistence of the slogan 'philosophy is analysis,'" Wilfred Sellars argues, "we now realize that the atomistic conception of philosophy is a snare and a delusion. For 'analysis' no longer connotes the definition of terms, but rather the clarification of the logical structure—in the broadest sense—of discourse, and discourse no longer appears as one plane parallel to another, but as a tangle of intersecting dimensions whose relations with one another and with extra-linguistic fact conform to no single or simple pattern" (1963: 171). The way out of such a tangle, as we have suggested, is to reconceive both empiricism and idealism in terms of the complexity of activity and historical events. Such a reconception subjects to critique the *simplicity* of science by suggesting the possibility that the "unity of science" is not necessarily a universal criterion of judgment. Davidson does this explicitly in his argument, and Hayles does it implicitly in reading the data of perceptual experiments against one another. In this argument, cognition or empiricism is neither the arbiter of science, as Rorty argues Kant had made it, nor is it dependent on the givenness of physical "reality," as Gibson argues when he defines "perception" as simply the apprehension of "events" that he describes unproblematically as "change in the layout of surfaces, change in the color and texture of surfaces, or change in the existence of surfaces" (cited by Gibson and Spelke 1983: 14). Such a reconception, moreover, also subjects to critique the very idea of *exhaustion* as well, suggesting, as Kuhn does in his example of "translations" of the same words, that the "same" givens—whether they be the "data" of the physical world or the "experiences" of phenomenology—inherit their "sameness" from changing contexts. If this suggestion is true, the exhaustion of phenomena (both "events" and "experiences") is unattainable. And finally, such a reconception questions the *generalization* of science in the ongoing "conversations" (or the lack of "conversation") Rorty describes—a category which itself, like "configuration," confuses activity and understanding—between people who inhabit different "paradigms," between those who argue, rationally and atomistically, and those who tell stories (1982: 92–93).

The conception of "knowledge" we are presenting is metonymic rather than hierarchic, eschewing the aestheticism of "simplicity," the reductiveness of "exhaustion," and hierarchy of "generalization." It calls for a "content" of what we are calling a version of narrative and a "method" of *multiplied narratives* (including the bare forms of reason embodied in Greimas's charts and arrows as species of narrative). But of course content and method defined in this way immediately suggest their own confusion. Bruno Latour presents this narrative method/content in his discussion of the relationship between the two traditions we have been describing. He emphasizes the phenomenology of signification in Continental semiotics and the physicality of facts in Anglo-American cognitive science in his description of the relationship between the visualization and the processes of cognition. With the term "visualization," Latour is describing the mechanisms by which the complexity of experience is reduced (in writing, print, and other means) to "information" which can be preserved and easily transported. He calls these mechanisms the "inscription" and "mobilization" of knowledge. "There are two ways," he writes,

> in which the visualization processes we are all interested in may be ignored; one is to grant to the scientific mind what should be granted to the hands, to the eyes and to the signs; the other is to focus exclusively on the signs *qua* signs, without considering the mobilization of which they are but the fine edge. All innovations in picture making, equations, communications, archives, documentation, instrumentation, argumentation, will be selected for or against depending on how they simultaneously affect either inscription or mobilization. (1986: 26)

In the first extreme, the activity of cognition—its *constitution* by means of the activity of its inscription—is replaced by an abstract agency of "mind." When Whitehead suggests that in "founding European philosophy and European mathematics," Pythagoras in "a flash of divine genius, [penetrated] to the inmost nature of things" (1967: 37), he is asserting that Pythagoras's "probing with cool dispassionateness into ultimate meanings" (1967: 33) was a function of the "human mind" (1967: 34).

In the second extreme, the social and cultural effects of cognition—its "mobilization" both in the sense that cognitive meanings can be transported from one social and historical situation to another and in the sense that cognition is mobilized within social and intellectual conflicts which are always interested—are replaced by the abstract functioning of signs. Greimas suggests that even ethics and ideology can be submitted to the logic of signification so that, "supposing that the main axiological models [within a semantic universe] . . . were described [and] . . . that the paradigms . . . and rules of transformation of the ideological models were sufficiently well known," semantic models "capable of bending individuals and collectivities towards new axiological structures" could be created (1983b: 160). In this notion he is asserting that the logic of semantics takes precedence over its social and cultural functioning. By focusing exclusively on signs *qua* signs, semiotics fails to acknowledge what Ricoeur calls "the fundamentally dialectical structure of the category of standing-for": "standing-for," he says, " . . . means by turns the reduction to the Same, the recognition of Otherness, and the analogizing of apprehension" (1988: 157).

These extreme tendencies, Latour argues, can themselves be "mobilized" by means of the juxtaposition of mobilization and inscription—by means of what he calls their "superimposition." The discrepancies between different scientific observations, he argues in tracing the importance of inscribed and transportable "visualizations" in the history of science, "proliferate not by looking at the sky, but by carefully superimposing columns of angles and azimuths. No contradiction, or counter predictions" he continues, "could ever have been visible. Contradiction . . . is neither a property of the mind, nor of the scientific method, but is a property of reading letters and signs inside new settings that focus attention on inscriptions alone" (1986: 20). The "new settings" Latour mentions are social and cultural situations, even intellectual settings. Science in particular and cognition in general progress by means of such *superimposition* of different "settings" and understandings upon one another. In a way we have already noted, the notion of superimposition (configuration) describes Greimas's semiotic square with all *its* arrows and shapes, but it does so only if we conceive of that square as exceeding its own logic to square the circle of self-evident truths—truths which are arbitrarily (but decisively) chosen to be placed in the first corner of the square in the first place. In another way, it also describes what Whitehead calls the cognitive "agency of human reason," whose chief function is to exhibit "connections between things," which "are extremely unobvious" (1967: 19). But here again, Whitehead's focus on the cognition of real but not readily apparent relations between things only exemplifies Latour's description of the superimposition of different cognitive understandings upon one another if we conceive of the agency of human reason mediated by inscriptions that transform "experience" into "cognition."

Cultural Cognition

The intention of *Culture and Cognition* is, in fact, to superimpose the understanding of knowledge and cognition implicit in Continental semiotics on that implicit in Anglo-American cognitive science to better understand the workings of cultural discourse. Moreover, we are attempting the superimposition of semiotics on cognitive psychology in order to describe the cultural institutions of understanding in our time. (An important element of our argument suggests that psychoanalysis has pursued similar aims.) Such an aim cannot ever be completed. The study of culture, Clifford Geertz has asserted, is "intrinsically incomplete" (1973: 29), just as Ricoeur argues, working out of the Continental tradition, that narrative discourse itself is "intrinsically incomplete" (1984: 144). These incompletions pursue the path of "an open-ended, incomplete, imperfect mediation, namely, the network of interweaving perspectives of the expectation of the future, the reception of the past, and the experience of the present, with no *Aufhebung* into a totality where reason in history and its reality would coincide" (Ricoeur 1988: 207).

In most of this book we examine different aspects of the complications and assumptions governing cognition. Here, however, we discuss the very nature of the "settings" (or "networks," as we have also called it and as we also describe the semiotic square) conditioning cognition. Those settings, we argue, are *cultural*. Not only do cultural assumptions about what is and is not "self-evident" constitute part of the network of cognition. Equally important is that cognition and acts of cognition help to constitute what we mean by "culture." Culture, we maintain, is a network of human activities—both cognitive and noncognitive—that emerges in time (see Ricoeur 1984: 196 for a discussion of this emergence), and this network is in a complex relationship to modes of knowing and formulations of value. In other words, cognition is not solely and fully conditioned by cultural and historical contexts. In Chapter 1, for instance, we are not arguing that because binary structures came to dominate models and modes of understanding in Western culture in the twentieth century, those structures somehow do not accurately (logically and precisely) describe and account for cognition. In more general terms, while Latour and the Edinburgh theorists argue that "'knowing' is not a disinterested cognitive activity" (Latour 1986: 16), knowledge can still accurately and functionally describe the world within particular settings that condition the canons of accuracy and functionality. This is part of what Felman means in her suggestion that cognition has to do with reality and what Samuel Wheeler means when he glosses his assertion that "routine categorization is . . . not a matter of just fitting the raw facts" by noting that this assertion does not "deny that there is a world, and that that world constrains what we say, and when," but it does deny "that our descriptions of that world are determined by that world exclusively" (1989: 135, 138).

Our intention then, is not to deny the possibility of cognitive "accuracy," but simply to ask why a mode of understanding that is, in fact, accurate emerges at a particular historical moment within a particular culture. Even Whitehead describes the specific cultural forces in Pythagorean Greece and in seventeenth-century Europe that allowed abstract mathematical thinking to condition science and philosophy (1967: 33–34). And even Greimas suggests that "a set of historical and pragmatic factors" created "a privileged place" for binary structures in understanding discourse and cognition in the twentieth century (Greimas and Courtés

1982: 25). In his argument, however, Whitehead positions those cultural forces as a kind of "background"; they allowed transcendental truth to be perceived simply because they were not as noisy and distracting as culture usually is. (In Chapter 3 we describe this kind of negative accommodation as "weak" preparedness for selection.) And in his brief suggestion, Greimas momentarily foregrounds the explanatory strength of formalism in science. But neither Whitehead nor Greimas, in these accounts, describes the intersection (or superimposition) of cognition and culture. Since they both imply that knowledge itself transcends the moment and mechanisms of its recognition, neither wishes to focus on the *emergence* of knowledge as a process that is governed by social and cultural institutions beyond the "background" accidents they both note. Neither practices what David Bloor and Clifford Geertz call "thick description," the representation of the multiplicity of conditions that contribute to the construction of emerging "truth." Geertz asserts that such description, like the "events" Davidson explores, rather than finding resolutions, constantly increases the tension "between the need to grasp and the need to analyze" (1973: 24). The very differences between Whitehead's analysis and Greimas's grasp—the differences in their assumptions about the nature of knowledge, its components, its sources, its development, and its deployment—are revealed in their superimposition. Moreover, that superimposition helps to demonstrate the institutional nature of cognitive activity, cognitive generalization conceived as historical events.

Before we can examine the possible institutional nature of cognition, we need to define more precisely what we mean by "culture." (What we mean by "cognition" will be defined and redefined in the three parts of this book.) "Culture" can be seen, among its other senses, to describe the relationships between global and atomistic conceptions that have been called paradigms. It is through the "concern with the conditions that make movement from local sites to global systems possible," as Hayles notes, that we expose "presuppositions within older paradigms that made universalization appear axiomatic" (1990: 16). In the last thirty years, two foundational usages of culture have developed, one associated with Raymond Williams and the other with Claude Lévi-Strauss. Their opposed versions of culture correspond to our opposition of the traditions of Anglo-American empiricism and Continental semiotics. In *Culture and Society* Williams traces the nineteenth-century articulation of culture as a concept and a setting for understanding. He offers a historical etymology of this concept, an etymology which can be conceived as a form of genealogy (see Davis and Schleifer 1991: 35–43), one that superimposes different definitions in a single text. In this way, Williams's study of culture is "empirical" rather that "rational." That is, instead of beginning with a transcendental definition of culture as an object of cognition that can be logically determined, Williams examines the historical and social uses of the term. He is pursuing the "mobilizations" of the concept throughout history rather than its "inscription" within what Edmund Leach, speaking of Lévi-Strauss, calls the "algebraic matrix of possible permutations and combinations located in the unconscious 'human mind'" (1970: 40). The procedure we are associating with Lévi-Strauss, Stuart Hall argues, is "the very opposite of a Hegelian search for underlying Essences." In Williams's conception of culture, however, Hall finds "common and homologous 'forms' underlying the most apparently differentiated areas" (1980: 64). This tendency makes Williams's work, at least in part, essentializing and, in this way, akin to Marx and Engels's belief, for example, that the historical dialectic they study is actually *in* nature. As in the Anglo-American empirical tradition more generally, this is a tendency toward the absolute opposition of form and content that formalism, rather than phenomenological structuralism, assumes.

Lévi-Strauss is a striking example of the semiotic tradition of "culture" because he wants to universalize cultural phenomena seemingly as attributes of signifying practices which lead, as we will see by way of psychoanalysis and Lacan, to conceptions of cognition focusing on discourse and the subject. This approach to cognition is close to the "inscriptive" practices of science—science as discursive notation, science as the formulation and recording of notations—that Latour and other ethnomethodologists describe. For Lévi-Strauss, discursive activity—especially narrative activity—provides structures of universal cognition, the particular apprehensions belonging to members of specific cultural settings. Here Lévi-Strauss's important distinction between structure and form, relationality and substance, is of the utmost importance. Lévi-Strauss wants to isolate universal, "rational" structures of cognition which will allow for a great variety of cognitive forms and, more important, the great variety of cultural forms. He wants to be able to account for universal and permanent cognitive inscriptions which, nevertheless, can manifest themselves in a great variety of cultures. This is a type of scientific enterprise, as Steve Woolgar notes, that applies scientific critique to its objects of inquiry but not quite to itself (1988: 21).

A similar confusion (or is it a "superimposition"?) of two definitions of culture can be found in Williams. Despite his seemingly empirical examination of culture, Williams makes Matthew Arnold's explicit definition of culture in *Culture and Anarchy* (1869) the touchstone of his study. In *Culture and Society* Williams quotes Arnold's famous definition of "culture as the great help out of our present difficulties; culture being a pursuit of our total perfection by means of getting to know, on all the matters which most concern us, the best which has been thought and said in the world; and, through this knowledge, turning a stream of fresh and free thought upon our stock notions and habits" (cited by Williams 1958: 115). Here culture is both a body of knowledge and a mode of behavior. Arnold sees the essential aspect of a definition of culture in its attempt to make "the best which has been thought and said in the world" prevail.

Arnold's global concept of culture, like that of Werner Jaeger in the twentieth century, eschews the atomistic sense of culture developed by anthropology (which Lévi-Strauss both uses and modifies). "Culture," Jaeger argues in *Paideia,* has come to possess the "trivial and general sense [of denoting] something inherent in every nation of the world, even the most primitive. We use it for the entire complex of all the ways and expressions of life which characterize any one nation. Thus the word has sunk to mean a simple anthropological concept, not a concept of value, a consciously pursued *ideal*" (1974: xvii). In a famous definition of culture, the anthropologist E. B. Tylor articulated in 1871 what came to be considered the first empirical or "scientific" use of the term Jaeger is describing. "Culture or Civilization," he wrote in *Primitive Cultures,* "taken in its wide ethnographic sense, is that complex whole which includes knowledge, belief, art, morals, law, custom, and any other capabilities and habits acquired by man as a member of society" (cited in Stocking 1968: 73). The difference between this definition and that of Arnold and Jaeger is that it creates, as Lévi-Strauss does, the possibility of a plurality of cultures: culture does not aim at the universal "perfection" and the adherence to the will of God Arnold describes in *Culture and Anarchy.* It is a relative and local concept—an empirical understanding—manifested as a particular culture at a particular moment. This opposition repeats the atomistic and global forms of understanding we have described, Geertz's "analysis" and "grasp."

This opposition, again, is neither absolute nor complete. In fact, George Stocking has argued quite persuasively that Tylor's concept of "culture" depends precisely on the currency of the less "scientific" description Arnold offers. Tylor's argument, as Stocking describes it, is fully the kind of cognitive "mobilization" of an idea in the *interested* polemical conflict of cognitive activity Latour describes as the social arena of knowledge. The cognitive achievements of science and technology, Latour says, come about "in an agonistic encounter between two authors." "Writing and imaging cannot by themselves explain the changes in our scientific societies, except insofar as *they help to make this agonistic situation more favorable*" (1986: 5). It is in such an *ideological* conflict with Arnold that Tylor chooses the word "culture" and links it to what, in Arnold, is its antithesis, "civilization." Tylor confronts Arnold precisely because he subscribes to the ideology of laissez-faire individualism that Arnold is attacking in *Culture and Anarchy*. From the vantage of such an ideology, Arnold's concept of "the best which has been thought and said in the world" does not make sense precisely because the ideology of laissez-faire individualism does not allow for culturally acknowledged standards. Tylor, then, redefines the framework of understanding—of cognitive activity—by mobilizing the concept of culture within a new context in which it is indistinguishable from that of civilization insofar as both are plural, both functions of historical accidents (Stocking 1968: 69–90). Here is an example of Kuhn's assertion that even the "same" word can be a "translation" of itself. In this analysis Stocking, like Williams, is offering a *critique* as well as an exposition of the concept of culture by superimposing one concept upon another. He is attempting, further, to historically *situate* what seems, in the Arnoldian tradition, to be a universal and transcendental "idea" and what equally seems, in the tradition of scientific anthropology, to be a self-evidently logical definition. (For a fuller discussion of "critique," see Davis and Schleifer 1991).

In a similar way, in the opposition between Williams and Lévi-Strauss we can see both the superimposition and the confusion of the concept of culture. The power of both Lévi-Strauss and Williams as thinkers lies in their ability to superimpose inscribed and mobilized conceptions of culture within very different cognitive frameworks. In each, that superimposition can be seen in key concepts. For Lévi-Strauss, the key to his rationalistic account of the inscribed structures of cognition is their mobilization across cultures. For Williams, the key to his more or less empirical history of "ideas" is the countervailing tendency in his etymological study—later made explicit in *Keywords*—that connects mobilized ideas of culture to their repeated inscription. Williams demonstrates the tendency of this distinction when he says that Arnold

> was caught between two worlds. He had admitted reason as the critic and destroyer of institutions, and so could not rest on the traditional society which nourished Burke. He had admitted reason—'human thought'—as the maker of institutions, and thus could not see the progress of civil society as the

> working of a divine intention. His way of thinking about institutions was in fact relativist, as indeed a reliance on 'the best that had been thought and written in the world' (and on that alone) must always be. Yet at the last moment he not only holds to this, but snatches also towards an absolute: and *both are Culture*. Culture became the final critic of institutions, and the process of replacement and betterment, yet it was also, at root, beyond institutions. This confusion of attachment was to be masked by the emphasis of a word. (1958: 128)

Here Williams is offering the kind of superimposition we are describing as both the result and cause of "cultural" activity. His understanding of Arnoldian culture participates in cultural critique, superimposing conceptions of culture as both the *fulfillment* and the *replacement* of existing institutions.

Such a superimposition marks the *complexity* of culture, just as the semiotic square marks the rational complexity of discourse. That is, Williams's critique is based on the *sociohistorical* investigation that we described elsewhere as "genealogical" (Davis and Schleifer 1991: 39, 152–57). Ricoeur, in fact, suggests that Nietzsche's genealogy offers a version of narrative comprehension. He situates genealogy between philology (a version of "phenomenology") and physiology (1988: 238) and claims that "a genetic evaluation of anything is at the same time an evaluation of culture" (1988: 236). Analogously, Michel Foucault understands Nietzsche's conception of genealogy as a mobilization of a "patiently documentary" enterprise that attempts "to record the singularity of events . . . in what we tend to feel is without history . . . not in order to trace the gradual curve of their evolution, but to isolate the different scenes where they engaged in different roles" (1977: 139–40). Cognition itself is one of those places or singular "events" that only *seems* to be without a history, a universal subfunction of human consciousness and instinct that takes place, in both Whitehead and Greimas, before a "background" of accidents. Latour's "superimpositions," Ricoeur's "configurations," Geertz's "cultural studies" all focus on the ways such "events" *emerge* from the background of "culture"—Ricoeur calls it a background "made up of the 'living imbrication' of every lived story with every other such story" (1984: 75)—in which they are entangled.

Woolgar shows that this "background" to the facts and objects of scientific inquiry is retrievable in a five-step narrative of the relationship of texts and so-called objects in scientific knowledge. (1) "Science" begins when an investigator merely writes down or somehow records potential relations and possible facts in a working document. (2) On this basis, the investigator is in a position to focus and clarify the potential object of inquiry through manipulation or revision within or among documents. (3) At the point at which the object of inquiry becomes well defined, the investigator easily, and perhaps "naturally," moves to posit the independent existence of the object being studied. (4) The investigator can and will clarify the inquiry further by elevating the object of inquiry to an existence both independent of and, eventually, even *prior* to the initial documents. From this point forward, the object of inquiry can be said to be only "reflected" or noted in the documents. (5) In the final stage of consolidating investigation and discovery, it is incumbent on the investigator—to remain consistent with empirical and scientific protocols—to deny the existence of steps (1), (2), and (3) (1988: 68).

Such a narrative analysis suggests that to position scientific discovery as a cultural formation, as we are attempting in this book, is to reintroduce a repressed historical consciousness—to mobilize a history—around and through that concept and activity. Doing so successfully (to the point of historicizing science itself) will inevitably create what Harold Garfinkel calls "reflexivity," a turning of scientific critique to investigate the very scientific instrument of that critique and will even create an inversion of the assumption that the object of inquiry precedes the documents that produced it (1967). This constructivist approach of ethnomethodology tends to erase the opposition between background and foreground, as Williams, Lévi-Strauss, and Woolgar do in their different ways—and, as we saw earlier, as Rorty, Davidson, Quine, and Sellars also do. Beginning with the universality of "mind," and then narrating the activity of "mind" in terms of a genealogy of concepts, Williams ends by describing multiple empirical sites of social and political conflict. Lévi-Strauss begins with the sociohistorical investigations of anthropology—the multiple sites of human cultures—and ends with the generalization of the matrix of inscriptions he describes in the functioning of "mind" and even in the chemistry of the brain. Woolgar, much as we are trying to do, works to retrieve the context and material "thickness" of such investigations of cognition as cultural activities.

The conception of culture that obviates the opposition between background and foreground creates an alternative to Whitehead's sense of the transcendental power of mind and Greimas's sense of the logic of combination and permutation that governs the generation of meaning within sign systems. In that understanding of culture we can see where the superimposition of the empirical tradition of Anglo-American conceptions of cognition and the rationalist tradition of Continental semiotics can suggest the precise *force* of science as a particular articulation of culture—in Stanley Aronowitz's phrase, "science as power" (1988). By deliberately staging aspects of this situation, this superimposition, we can then identify various *institutions* of understanding by focusing on narration as Lévi-Strauss and Williams do in different ways.

A concrete example of retrieving the institutional nature of understanding—the final example in this Introduction that superimposes in its own argument general cognition and particular history—is Stephen Jay Gould's history of the study and measurement of "intelligence" in *The Mismeasure of Man*. Like Woolgar, Gould is attempting to criticize "the myth that science itself is an objective enterprise, done properly only when scientists can shuck the constraints of their culture and view the world as it really is" (1981: 21). Instead, Gould argues that science is a social phenomenon. The science he traces is what Charles Spearman called the "science of cognition" (cited in Gould 1981: 262). That science attempted to isolate intelligence as a "single quantity" so that it could be measured (1981: 20), in much the same way Gibson attempted to isolate perceptual "events" as single "changes" in the world that could be measured. Gould traces the history of intelligence measurement and testing in rich detail to describe the culturally determined racism and sexism that leads to what he calls "*unconscious* finagling" (1981: 54). "Science," Gould argues, "is rooted in creative interpretation. Numbers suggest, constrain, and refute; they do not, by themselves, specify the content of scientific theories. Theories are built upon the interpretation of numbers, and interpreters are often trapped by their own rhetoric. They believe in their own objectivity, and fail to discern the prejudice that leads them to one interpretation among many consistent with their numbers" (1981: 74). In the specific case of testing and measuring "intelligence," Gould argues that recognizing "the importance of mentality in our lives" and wishing to characterize it ("in part so that we can make the divisions and distinctions among people that our cultural and political systems dictate"), we give "the word 'intelligence' to this wondrously complex and multifaceted set of human capabilities" (1981: 24).

The reduction Gould is describing encompasses the five steps Woolgar presents in the sociology of science, and one could outline Gould's detailed history of the "science of cognition" in Woolgar's abstract five-step analysis. More important for our study, however, is Gould's discussion of what he calls "cultural evolution" at the end of his study. If "the evolutionary unity of humans with all other organisms is the cardinal message of Darwin's revolution" in science, this doesn't reduce the human to "'nothing but' an animal" within a framework of empirical positivism. Such an assertion, Gould argues, is just as fallacious as the description of the human "as 'created in God's own image'" (1981: 324) within a framework of rational idealism. Rather, human life "has established a new kind of evolution to support the transmission across generations of learned knowledge and behavior," what Gould defines as "cultural evolution" (1981: 324–25). He calls this Lamarckian insofar as it allows for the inheritance of *acquired* traits. The concept of cultural evolution allows him a narrative in which "our large brain is the biological foundation of intelligence; intelligence is the ground of culture; and cultural transmission builds a new mode of evolution more effective than Darwinian processes in its limited realm—the 'inheritance' and modification of learned behavior" (1981: 325). In Part I we explore a structure of such "transmission" by surveying empirical studies of the cognition of old people in hopes of presenting a version of Rorty's "epistemological behaviorism."

Within a more traditional Darwinian framework Clifford Geertz makes a parallel argument: he offers a similar narrative description of human cognitive development. Geertz argues that "the human brain is thoroughly dependent upon cultural resources for its very operation; and those resources are, consequently, not adjuncts to, but constituents of, mental activity" (1973: 76). Because of this dependence, Geertz argues that thinking and cognition do not consist "of 'happenings in the head'" but rather of "public" *cultural* activities (1973: 45, 12). For Geertz, the cultural component of cognitive evolution is Darwinian as well as Lamarckian: he argues that "not only was cultural accumulation under way well before organic development ceased, but that such accumulation very likely played an active role in shaping the final stages of that development" (1973: 67). That is, he argues that the developing nervous system adapted to the emerging cultural environment so that "culture, rather than being added on, so to speak, to a finished or virtually finished animal, was ingredient, and centrally ingredient, in the production of that animal itself" (1973: 47). Consequently, he says, "the standard procedure of treating biological, social, and cultural parameters serially—the first being taken as primary to the second, and the second to the third—is ill-advised" (1973: 74).

Gould and Geertz argue that the opposition between fact and reason—the empiricism and rationalism that has governed the exposition of this Introduction and of "culture" and "cognition" more generally—creates a false absolute and betrays, so to speak, true middles. Such middles are composed of Gould's "wondrously complex and multifaceted set of human capabilities" (1981: 24) and Geertz's "dispositions." "The term 'mind,'" Geertz writes,

> refers to a certain set of dispositions of an organism. The ability to count is a mental characteristic; so is chronic cheerfulness; so also—though it has not been possible to discuss the problem of motivation here—is greed. The problem of the evolution of mind is, therefore, neither a false issue generated by a misconceived metaphysic, nor one of discovering at which point in the history of life an invisible anima was superadded to organic material. It is a matter of tracing the development of certain sorts of abilities, capacities, tendencies, and propensities in organisms and delineating the factors or types of factors upon which the existence of such characteristics depends. (1973: 82)

Above all, these factors include the complex and multifaceted set of human capabilities—in Ricoeur's term, the *configurations* of capabilities—we call culture. Versions of such "culture" are Rorty's conversation, Nietzsche's genealogy, even Gibson's positive empirical science and Greimas's binary rationalism. They all present narratives that to one extent or another create versions of cognition and understanding. These narratives can be articulated and entangled—they can be superimposed upon one another, as we are trying to do here—but they can never quite be reduced to one or another master narrative.

In the following chapters we multiply and superimpose such narratives. In Part I we pursue the simplicity, generalization, and exhaustion of understanding and cognition in three chapters that examine narrative structures of cognition—binary oppositions, the general form of narrative discourse, and the uses of narrative for the transmission of cultural values. In Part II we explore the same categories of empirical scope, cognitive simplicity, and the generalizing understanding of rhetoric in particular cases of cognition. Finally, in Part III we discuss the institution of knowledge in the activities of teaching, dissemination, and reading—all forms of cultural discourse—which again follow, in a general way, the exhaustion of data, the generalization of observations, and the simplicity of analysis that comprise a method of examining cognition and culture taken together.

Part I

NARRATIVE STRUCTURES

I

COGNITION AND NARRATION

Binary Structures and the Nature of Meaning

Cognitive Modeling

In *Semiotics and Language,* A. J. Greimas and Joseph Courtés argue that "a set of historical and pragmatic factors has given binary structures a privileged place in linguistic methodology." "This may be due," they say, "to the successful practice of the binary coupling of phonological oppositions established by the Prague School, or due to the importance gained by binary arithmetical systems (0/1) in automatic calculus, or to the operative simplicity of binary analysis in comparison with more complex structures, since every complex structure can be formally represented in the guise of hierarchy of binary structures, etc." (1982: 25). Greimas's semiotic square is one such hierarchy of binary structures, mapping what he calls "the elementary structure of signification." That structure, Greimas argues, is one by means of which "the human mind" constructs cultural objects "out of a desire for intelligibility" (1987: 48). In the next chapter we examine this elementary structure in relation to the intelligibility of narrative discourses, and in Chapter 5 we examine the desire for intelligibility itself in relation to psychoanalysis. Here, however, we want to examine the simple abstract modes of intelligibility and cognition itself and the *situations* of twentieth-century culture that gave rise to the understanding of that abstraction.

Cognition and cognitive science, as they have developed as concepts and programs for research in the context of the pragmatic and historical factors of twentieth-century science and intellectual life, are based on the assumption that intelligibility functions by means of generalizing simplicity that is not contradicted by experience. Three elements of this formulation are essential to understanding. (1) Understanding abstracts from experience in order to discover or articulate invariants within the complexity of phenomena, invariants which can be *generalized* across a host of particular "facts" and phenomena. (2) Moreover, it does so in ways that are internally consistent (i.e., not self-contradictory) and parsimonious, attempting to achieve its understanding by the *simplest* means necessary. (3) It also does so in ways that account for as much empirical data as it can. Different approaches to understanding can emphasize one or the other of these elements: in twentieth-century linguistics, for example, Bloomfieldian structuralism emphasized the empirically *exhaustive* nature of understanding, choosing the "factual" range as its chief goal; Louis Hjelmslev and the Copenhagen school of linguistics emphasized the logically consistent nature of understanding, choosing the *simple* systematicity of understanding as its chief goal; and Roman Jakobson and the Prague school emphasized the *generalizing* abstract nature of understanding, choosing the articulation of the invariables of language as its chief goal (see Schleifer 1987a: 44–66). In these terms, Greimas is probably closest to Hjelmslev—he called the *Prolegomena to a Theory of Language* "the most beautiful linguistic text" he had ever read (1974: 58)—yet in his work, as in the various disciplines associated with cognitive science more generally, all three of these elements are always present to some degree.

In other words, cognition itself, as a concept and an object of scientific inquiry in the twentieth century, has been understood to be the simplification and generalization of experience. In fact, Robert Rubinstein, Charles Laughlin, and John McManus—anthropologists and psychologists working in an Anglo-American empiricist tradition whose assumptions about what constitutes specific "knowledge" are very different from those of Greimas and the rationalist tradition of Continental linguistics and semiotics—assert in *Science as Cognitive Process* that "perhaps the major function of the brain is to model reality" (1984: 21). That is, the nature of cognition is to simplify experience. Such modeling is a mode of simplification that can be understood in many ways. It "reduces" uncertainty; it filters out irrelevant data; and it allows for the immediate (phenomenal) apprehension of Greimas's elementary structure of signification. (Here again we are repeating the elements of simplification, generalization, and exhaustive accounting in cognition.) "Cognition," Rubinstein and his colleagues say, "is a simplified representation of the operational environment" (1984: 26), an environment of "events." "Although the [linguistic] message is presented for reception as an articulated succession of significations," Greimas writes, " . . . reception can be effectuated only by transforming the succession into simultaneity" (1983b: 144)—which is to say, transforming the complexity of experience into simple understanding. Greimas's usual term for "reception" (at least in his earlier work where he is following Claude Lévi-Strauss) is "apprehension," and in *Semiotics and Language* he and Courtés describe cognition as a "specifying term" in relation to the "reception" of knowledge (1982: 32).

Simplification is usually taken to mean a mathematical reduction to common denominators. But another form of simplification is more idiosyncratic: the reduction to the already known, the repression of irregularities—what Simone de Beauvoir calls the positive value of habit (1973: 696). This is the difference between simplifying cognition and a more habitual, comfortable cognition. We could also describe it as the difference between logical and ecological cognition. In an important way this distinction is a significant contribution of feminism to the understanding of cognition, an understanding that seeks to emphasize what Mary Hawkesworth calls "the conception of cognition as a human practice" (1989: 551). It seeks to emphasize the ecology of cognition. "Feminist objectivity," Donna Haraway argues, "means quite simply *situated knowledges*"; it is "a doctrine and practice of objectivity that privilege contestation, deconstruction, passionate construction, webbed connections, and hoped for transformation of systems of knowledge and ways of seeing. But not just any partial perspective will do; we must be hostile to easy relativism and holisms built out of summing and subsuming parts" (1991: 188, 191–92). Comfortable cognition, then, does not mean "easy" cognition so much as it means "situated" cognition.

Such a conception, Hawkesworth writes, examines "the specific processes by which knowledge has been constituted within determinate traditions," and it also focuses on "the theoretical constitution of the empirical realm" and illuminates "the presuppositions that circumscribe what is believed to exist and identify the mechanisms by which facticity is accredited and rendered unproblematic" (1989: 551–52). We are calling such cognition "comfortable" because instead of simplifying in a hierarchical system, the practical functioning of such cognition emphasizes what Sandra Harding calls the "emotional labor or the positive aspects of 'relational' personality structures" that are encountered "when we begin inquiries with women's experiences instead of men's" (1986: 446). "Comfort" suggests the critical relation between "ordinary" experience—the "webbed connection," hope, and passion Haraway describes—and its contextualization as a practice.

As we will see in the case of old people, such an emphasis situates cognitive activity within the greater ecology of human life. In the tradition of Continental semiotics Julia Kristeva emphasizes this "ecological" understanding of cognition in her description of the "transcendence" of reason by the "heterogeneous element" of life that challenges "the speaker with the fact that he is not whole, but [it does] so in a manner altogether different from that in which the obsessed person's wretched consciousness ceaselessly signifies his bondage to death. For if death is the Other, life is a third party; and as this signification, asserted by the child is disquieting, it might well unsettle the speaker's paranoid enclosure" (1980: 271). Kristeva, emphasizing as she does the familial and maternal, prompts our opposition between "logical" and "comfortable" cognition, the opposition (but also the relation) between cognition and narration we are pursuing throughout this book. She accomplishes this emphasis by turning, as she claims Rousseau and Freud did, to childhood—the "third party" mentioned above—to find a narrative that "joins cause and effect, origin and becoming, space and time" (1980: 275). But it is also possible to turn to the elder rather than the child to discover a narrative of cognition that emphasizes the nonhierarchical ecology of cognitive activity.

This emphasis can be seen more fully in other feminist critiques of science that attempt to redefine cognition outside the semiotic tradition within which Kristeva works. Evelyn Fox Keller, for instance, distinguishes between "order" and "law" and suggests the opposition between comfortable and logical cognition we are pursuing here. "The concept of order," she writes, "wider than law and free from its coercive, hierarchical, and centralizing implications, has the potential to expand our conception of science. Order is a category comprising patterns of organization that can be spontaneous, self-generated, *or* externally imposed; it is a larger category than law precisely to the extent that law implies external constraint" (1985: 132). "The concept of nature as orderly, and not merely law bound," she concludes, "allows nature itself to be generative and resourceful—more complex and abundant than we can either describe or prescribe" (1985: 134). Such a concept allows Keller to distinguish between difference and dichotomy: "difference," she writes, "constitutes a principle for ordering the world radically unlike the principle of division or dichotomization (subject-object, mind-matter, feeling-reason, disorder-law)" (1985: 163)—in a manner directly parallel to Jacques Derrida's critique of semiotics and structuralism in terms of the *difference* between conceptions of difference and opposition (1979: 149; see Davis and Schleifer 1991: 168–72).

Derrida examines difference and opposition in terms of the dichotomization of sexual difference, but Keller examines these concepts more closely in relation to science and cognition. Specifically, her example is the work in genetics of Barbara McClintock and the difference between *simplifying* hierarchical structures of knowledge in biology and a conception of "steady state." This opposition is "the contest that has raged throughout this century between organismic and particulate views of cellular organization—between what might be described as hierarchical and nonhierarchical theories" (1982: 124). In the former, a "master molecule"—whether it be the cell nucleus or DNA—is described as the simplifying "center" or "origin" of a "linear hierarchy." The latter concept, exemplified by McClintock's research, "yielded a view of the DNA in delicate interaction with the cellular environment—an organismic view" (1982: 125). In this work, Keller argues, "no longer is a master control to be found in a single component of the cell; rather, control resides in the complex interactions of the entire system" (1982: 125). Such a view has profound implications for the study of cognitive activity within the ecology of human life. Instead of taking one "component" of the species (such as a male) or one moment of human development (such as adulthood) as the simplifying single "representative" of humanity or of what Keller elsewhere calls "cognitive maturity" (1985: 84)—a "representative" against which other components are measured—the necessity for cognitive simplification itself can be subject to critique. This critique of cognition, which we develop later, is essentially cultural. (For a parallel discussion in terms of the rhetoric of the simplifying hierarchy of synecdoche and the complexity of metonymy, see Schleifer 1990: esp. chap. 1.)

For both Continental and Anglo-American researchers, however—even across disciplines as different as psychology, anthropology, linguistics, neuroscience, philosophy, and computer science—cognition has been conceived as a way of simplifying and reducing experience to "knowledge" that, to some degree, is detachable from the experience out of which it arises, knowledge that is generalizable. In all these disciplines, moreover, the *method* of such generalization has assumed the form of the "binarity" which Greimas and Courtés describe as privileged in the twentieth century—a form which assumes that difference is most simply conceived as opposition or dichotomy. In fact, the key models that have been used to articulate the processes of cognition in the twentieth century have taken binary form in one way or another. In *The Mind's New Science*, Howard Gardner describes different models used to describe cognitive activity, all of which are based on the assumption that cognition is a cross-cultural phenomenon. Supporting this assumption, Gardner argues, is evidence from studies of syndromes resulting from casualties in the great wars of this century, pointing to very similar symptoms despite vastly different cultural contexts. This finding suggests basic, underlying processes and structures of thought—generalizable processes—which can be modeled.

The first model for the processes of cognition Gardner describes is that of "mathematics and computation." This model is based on the logical-mathematical work of Bertrand Russell and Alfred North Whitehead that ultimately resulted in the theorem of Alan Turing which demonstrated that a simple binary machine "could in principle carry out any possible conceivable calculation" (1985: 17). Gardner also presents the "neuronal" model, which describes the activation of neurons in terms of binary oppositions very similar to Turing's model. A model for cognitive activity closely related to both of these is that of "information theory," in which, as in the models of the computer and of neurons, binarity is of utmost importance in that information theory is based on the understanding that binary oppositions (0/1) can be used to describe information "in a way entirely divorced from specific content or subject matter" (1985: 21). In cognitive science these models can be taken to be rigorously isomorphic with the functioning of human cognition: Turing's description of any conceivable calculation is a simplifying model; and information theory presents a model that is generalizing in structuring "information" without recourse to the "content" of that information; and the neuronal model asserts the empirical reality of binarity in the functioning of the brain.

Thus, even in this description of cognition far removed from the terms and understanding of Greimas and the Continental semiotic tradition, cognitive science in the twentieth century—models of understanding and thought in our time—is based on the "privileged place" of binary structures in mathematics and computation (where "every complex structure can be formally represented in the guise of hierarchy of binary structures"), in information theory (as evidenced in "the successful practice of the binary coupling of phonological oppositions established by the Prague School"), and in the "automatic calculus" of computer science (Greimas and Courtés 1982: 25). All these binary models, as Gardner says, were "based on a shrewd hunch: that human thought would turn out to resemble in significant respects the operations of the computer, and particularly the electronic serial digital [i.e., "binary"] computer which was becoming widespread in the middle of the century" (1985: 43–44). This "hunch" is also related to what Gardner calls "the cybernetic synthesis" of "self-correcting and self-regulating systems." Such systems, in which "the human nervous system, the electronic computer, and the operation of other machines" could be understood as analogous, were named "cybernetic" systems (1985: 20–21), and they were understood to be analogous to one another—and to cognition in general—in a gesture of comprehension that is generalizing, simplifying, and accurate.

Metaphorical Cognition

One example of the ways in which cognitive science has attempted to understand the functioning of understanding can be seen in recent work in cognitive psychology to devise experimental methods for measuring human understanding. In the cognitive sciences recent discussion has focused on what F. C. Bartlett calls the "effort after meaning" (1932) and the processes of comprehension. One locus for such studies is the functioning of metaphors in language, which from classical times has been used to describe semantic cognition in terms of binary oppositions. Jonathan Culler discusses the curious status of metaphors within literary studies in recent criticism that sheds light on the study of metaphoricity as a cognitive process. Culler suggests that metaphor is not simply one figure or trope among the array of tropes, but rather "the figure of figures, a figure for figurality" (1981: 189). More recently Stuart Peterfreund, citing Culler, has argued that the figure "used by those aspiring to certain knowledge [and] . . . to empowerment through the attainment of certain knowledge" is metonymy (1991: 28; see also 1990a: 141–42). This is best seen, Peterfreund says, in "the function of metonymic logic in Newtonian thought [which] is justificational as well as explanatory, theological as well as scientific" (1991: 29). In this argument Peterfreund is unpacking—or, as Greimas might say, "exploding"—the simple concept of "metaphor" by demonstrating its complex functioning within understanding. He does this by showing the cultural category of empowerment in the "scientific" use of causal (i.e., metonymic) explanation and by emphasizing, with Umberto Eco, the necessity of distinguishing metonymy and synecdoche as well as metaphor and metonymy (see Eco 1976: 281).

Whereas for Peterfreund, the functioning of metaphor calls for the analysis of its rhetorical and philosophical functioning in Aristotle, Newton, Eco and others—after all, in a series of articles Peterfreund has pursued rhetorical analyses of the relation between Newtonian science and Romantic poetry (1990a; 1990b; 1991; see also Bono 1990)—for cognitive science, such a description calls for experimental examination, reproducible experiments that aim at accuracy and simplicity and that do not contradict a general theory of cognition. Such experimentation takes place at the juncture between a conception of scientific knowledge that conceives of such knowledge as "objective" truths, verifiable by replicable experiment, and a second conception of knowledge that conceives of knowledge itself as interpretations rather than (or along with) "truth"—with what Barry Barnes calls the "inalienable social, collective dimension" in the constitution of knowledge, "including scientific knowledge" (1983: 20). Cognition, Barnes argues (once again in terms of binary oppositions), is the recognition of similarities—what we might call the invariants within the variants of experience. "An assertion of resemblance," Barnes writes,

> which is what the application of a concept amounts to in this case, involves asserting that similarities outweigh differences. But there is no scale for the weighting of similarity against difference given in the nature of external reality, or inherent in the nature of the mind. An agent could as well assert insufficient resemblance and withhold application of the concept as far as reality or reason is concerned. . . . All applications of [the concept] 'dog' involve the contingent *judgment* that similarity outweighs difference in that [particular] case. (1983: 26)

Working in the Continental tradition of semiotics, Greimas describes the perception—the re-cognition—of language in remarkably similar terms. "We perceive differences," he writes, "and thanks to that perception, the world 'takes form' in front of us and for us." Such perception, he goes on to say, is a function of both difference and similarity: "to perceive differences means to grasp the relationship between the terms, to link them together somehow" (1983b: 19).

For both Barnes and Greimas, then, cognition "involves" the binary relationship between similarity and difference. In both, we can see the functioning of binarity in the description of understanding. This is why we can best give an example of cognition here by discussing the psychological study of metaphors within the Anglo-American tradition of cognitive science. (In Continental semiotics, which does not assume the self-evident preexistence of objective facts in relation to their perception, the obviousness of the opposition between literal and figurative articulation is not so clear.) Metaphor in both Culler's global description and Peterfreund's atomistic description is, as Aristotle describes it in the *Poetics,* the locus of similarity and difference: it consists, Aristotle says, in giving a thing a name that belongs to something else "by transference either from genus to species, or from species to genus, or from the species to species, or by analogy, that is, proportion" (1989: 76). As such, metaphor is the locus of intelligence, the very locus of cognition. Such transference, as Ricoeur (among many others) has argued, is not simply substitution: metaphors *augment* understanding "with meanings that themselves depend upon the virtues of abbreviation, saturation, and culmination, so strikingly illustrated by emplotment" (1984: 80). In *The Mill on the Floss*, George Eliot narrates the knowledge and power inherent in metaphor globally and atomistically conceived.

> It is astonishing what a different result one gets by changing the metaphor! . . . It was doubtless an ingenious idea to call the camel the ship of the desert, but it would hardly lead one far in training that useful beast. O Aristotle! If you had had the advantage of being "the freshest modern" instead of the greatest ancient, would you not have mingled your praise of metaphorical speech, as a sign of high intelligence, with a lamentation that intelligence so rarely shows itself in speech without metaphor,—that we can so seldom declare what a thing is, except by saying it is something else? (1961: 124)

Earl Mac Cormac has devoted a full-length study to the functioning of metaphor as a cognitive process. Metaphors, he argues, "can be described as a process in two senses: (1) as a cognitive process by which new concepts are expressed and suggested, and (2) as a cultural process by which language itself changes" (1985: 5–6). In this, Mac Cormac is arguing that metaphors are *mobilized* in the processes of understanding even while they are *inscribed* within the more or less unconscious cultural-linguistic definitions of understanding. That is, he combines the metonymic mobilization—empirically across data and philologically across cultural history—that Peterfreund describes and the figural inscriptions Culler asserts in calling metaphorical substitution a figure of figures, a figure for figurality.

Understanding and Empiricism

It is not our aim here to offer an exhaustive examination of the functioning of metaphor. Rather, we want to describe the empirical tradition of experimental psychology. In that tradition, measurement of empirical "data" is of the utmost importance: it is assumed that the index of effort one puts into cognitive activity can be measured (in carefully constrained situations) with reaction data. Thus, the time a person takes to interpret some bit of language is a window onto the effort of understanding: the transparency of the window is determined by the refinement of research manipulations. Cognitive psychologists use specific research paradigms to produce their data. While cognitive psychologists are trained to let the empirical result "speak for itself" by using objective tools of analysis such as inferential statistical analyses of numerical data, these psychologists inevitably interpret the refined behavioral data just as critics interpret literary texts—or, in Barnes's model, the subject of knowledge (whether individual or collective) "applies" concepts to phenomena through a process that involve contingent judgments. In the cognitive sciences a significant unresolved issue involves the effort required to comprehend a metaphor, the effort of understanding. Most cognitive psychologists assume that a person has limited capacity or resources to process information (Hirst and Kalmar 1987) and that we try to conserve these resources by using them efficiently (Navon and Gopher 1979). We can see in these assumptions the simplifying and generalizing aims of science in the specific injunction toward efficiency and a general "law" of the conservation of energy.

Some philosophers, like Donald Davidson, assume that metaphor is directly and easily comprehended because all the meaning is contained within the words and their grammatical context (1978: see Wheeler 1989: 119). Others, such as John Searle, argue that we interpret metaphorically only when language does not make literal (i.e., obvious) sense (1979). In this, he suggests, the nature of metaphor is local and situated: for instance, the sentence "some birds are carrots" is nonsense in the abstract and in most situations, but it is a meaningful figure in the context of orange birds (see Percy 1979: 64–82). At the farthest extreme, Nietzsche argues that all meaning is figurative, that it is the imputation of significance—which is to say, the "inscription" of cognitive structures—upon more or less undifferentiated experience. Barnes in his discussion of concept application comes close to this position.

Cognitive psychology attempts to measure the functioning of metaphorical comprehension, even though quite often the very experimental design of such measurements silently entails assumptions about the nature of the activity it measures. Samuel Glucksberg, for instance, uses a sentence verification paradigm to test whether people can ignore the metaphorical meaning of statements when they are asked to quickly judge literal truth or falseness of many sentences—some of which are powerfully metaphorical (Glucksberg et al. 1982; see also Gildea and Glucksberg 1983). In this paradigm, Glucksberg assumes that cognition is simply a matter of recognizing matters of fact (true or false). In discussing his experiment, he argues that when a metaphor such as "Some jobs are jails" is encountered during this true/false reaction-time task, subjects want to respond "false" but hesitate because they cannot ignore available metaphoric interpretations. Glucksberg's data seem to support his thesis. In his experiments, reaction time to deny the literal truth value of metaphors is slower than reaction time to deny what psychologists call "standard" false assertions (e.g., "some birds are carrots")—assertions which, in the abstract, are "obviously" false. (In such assertions of "obviousness" we can see the element of self-evident truths: what makes such propositions "obvious" in their truth value is the manner in which the sentence is detached from all social and interpersonal contexts. For a discussion of such abstraction, see Bakhtin 1986a: 99–106.) Glucksberg calls this discrepancy between the speed with which a subject judges an obviously false abstraction "false" and the speed with which the same subject judges an abstraction that can be figuratively true "false" a metaphor interference effect. These data can be interpreted to mean that literal and metaphorical meaning are accessed—that is, cognitively processed—simultaneously. In this interpretation, the recognition of metaphorical meaning does not merely follow *after* a failed attempt to interpret a sentence literally, as Searle has suggested. Rather, only when subjects are enjoined to measure metaphorical abstractions against abstract standards of "objective" literal truth (in terms of the binary opposition true/false), does the time of cognitive processing take longer. This suggests that the binary opposition between the "objectively" literal and the "nonobjective" figurative presents problems: the *meaning* of figures is processed—recognized—as quickly as so-called objective, literal meaning.

Another cognitive psychologist, Anthony Ortony, has studied the asymmetry of metaphors, the fact of their irreversibility (1979). While "a friend is an anchor" seems an acceptable metaphor to subjects in an experimental situation, "an anchor is a friend" seems odd to the same subjects. Here, in the cognition of metaphorical meaning, the relationship between similarity and difference—which functions in all cognition—appears as a special case. Metaphors contrast a low salient attribute of the grammatical subject with the same but now high salient attribute of the predicate. In this example, for instance, stability is a valid attribute but not a highly salient attribute of the concept "friend," while stability is a defining, highly salient attribute of "anchor." In his work, Ortony had people rate the quality of metaphors in both the standard and reversed orders of subjects and predicates. His experimental subjects clearly preferred the standard order of presentation in their ratings, but he did not record response speed to assess the difficulty of interpreting reversed metaphor.

The difficulty of separating a metaphor's effectiveness from a metaphor's interpretability is at issue. This is the difficulty of separating the articulation (or generation) of meaning from the communication of meaning, an issue of the utmost importance to Greimas and to Continental semiotics (see Parret 1983; Schleifer 1987a: chap. 5). But, as we have seen, it is also of great importance to Mac Cormac, and it is central to Peterfreund's distinction between figures that function by substitution (metaphor) and figures that join justification and explanation, physics and metaphysics (metonymy). The questions left unanswered by Ortony—and by the very "empirical" assumptions embedded in the use of time measurements to study the cognitive functioning of metaphors—are the following: Are the best metaphors, understood in the cognitive functioning of the articulation of meaning, easily interpreted, or are they interpreted at considerable effort? Is the effort of interpretation, whether easy or time-consuming, related to cognitive articulation? Moreover, if the "best" metaphors involve more effortful cognitive processing (in either articulation or interpretation), does this "effort" involve more information—situated rather than abstract "objective" propositions within an ecological rather than a logical system, what linguists call "marked" rather than "unmarked" forms (see Schleifer 1991: 293–96)? And, finally, can "meaning" be distinguished from "truth" precisely by means of this cognitive effort?

In a series of experiments to assess these questions, Monica Gregory and Nancy Mergler used response time as an index of cognitive effort. A few changes in the Glucksberg paradigm allowed for this. Glucksberg had placed his subjects in a *literal truth response set* in which they responded to a sentence's literal truth or falsity. A typically "true" sentence is "some fruits are oranges"; a "standard false" sentence is "some fruits are robins" (Gregory and Mergler 1990: 158). This is a fairly standard experimental manipulation, but it is a strange way of dealing with language in that it skirts the issue of meaningfulness. Gregory and Mergler, however, added a *metaphoric response set* to the Glucksberg paradigm so that subjects were asked to respond "yes" to sentences that seemed to present "meaningful metaphors" and "no" to sentences that seemed to present no metaphorical meaning. In these situations, subjects could follow the instructions and respond "no" to literally true sentences when no metaphorical meaning was apparent. Moreover, the error rates were the same as those in the literal truth response set. In these experiments, as in Glucksberg, subjects responded faster in judging ("true" or "false") literal truth than in judging metaphorical meaningfulness, which suggests that this second judgment is harder and more effortful. This finding is quite important, because it suggests that subjects can separate "truth" and "meaning."

In these two sets of experiments the same sentences were presented to subjects, and they were able to judge these sentences "false" in the first set (Glucksberg metaphor interference effect) and "meaningful metaphors" in the second, and also "true" and "nonmetaphorical." This is striking because in the second instructional set there could be no "errors" as such: as Barnes suggests, a judgment of metaphoricity can always be made. Still, the sentences most frequently judged metaphorical were standard metaphors (e.g., "some billboards are warts") and reversed metaphors (e.g., "some warts are billboards"), both of which—like all the figurative language in the Gregory/Mergler experiments—are Aristotle's species/genus relationships. Less likely to be judged metaphorical by these subjects were scrambled metaphors (e.g., "some billboards are princesses") and "standard false" propositions. But any utterance can yield metaphorical interpretation, even the Gregory/Mergler example of a "standard false" proposition, "some birds are mangoes"—as Barnes says, in a *particular* situation "there is no scale for the weighting of similarity against difference given in the nature of external reality, or inherent in the nature of mind" (1983: 26)—even though it takes great effort (more time) to interpret standard false sentences as good, meaningful metaphors. That is, in these cases more information needs to be processed.

The data from these experiments suggest that the easiest task for subjects was to say "true" to literal truth and then to say "no" to the possibility of metaphoricity in scrambled metaphor and standard false sentences (to judge a sentence to be nonsensical). Saying "no" to the metaphorical possibility of literal truth was third fastest. Here, the potential interpretation of meaning in a sentence that is not true slowed down cognition. This is what Walker Percy calls the phenomenon of "mistakes—misnamings, misunderstandings, or misremembering [—resulting] in an authentic poetic experience" (1979: 65), a cognitive experience of meaning. But even though it took more time, this, too, was a judgment of the nonsense of the sentence. Finally, the slowest and most ambiguous response was to say "yes" to metaphoric meaningfulness.

Gregory and Mergler argue that language decisions for ordinary language comprehension entail these three stages of judgment—three binary cognitive activities—because such staging conserves "effort after meaning" in the long run. First, they argue, the subject of linguistic cognition scans for literal, obvious *truth* (as opposed to untruth); then it rejects information as *nonsense* (as opposed to meaning); then it seeks "hidden," metaphorical meaning, precisely because the effort it requires entails the *marking* of information with the addition of more information. Such marking is the contextualizing and re-contextualizing of meaning—the very *situating* of meaning in relation to the "objective" truths of the world. These stages in the process of the cognitive apprehension of metaphors can be drawn on Greimas's semiotic square.

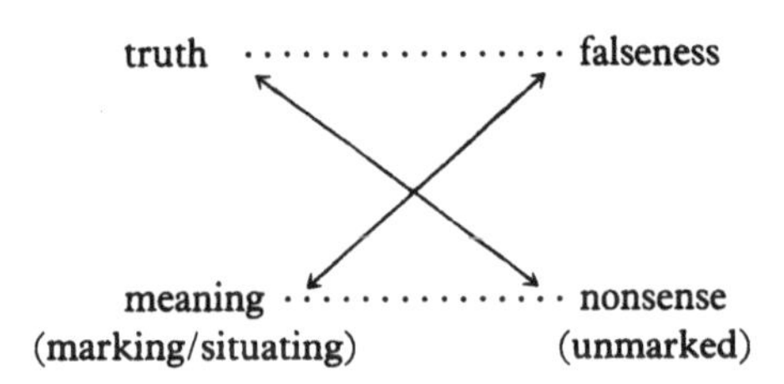

This schema maps the cognitive activity of semantic processing. It does so by describing the absence of truth not as "falseness" but as undifferentiated, unmarked "nonsense." This suggests that meaning and truth are not identical—their identity is an assumption in Whitehead's description of thought's harmony with reality and in Glucksberg's original experimental design—and it suggests that the mark of their difference is the *process* of generating meaning, the marking and situating of meaning. These conceptual syncretizations—"unpacking" the opposite of /truth/ as /falseness + nonsense/ and "unpacking" the complexity of /meaning/ itself as /marked + situated/—are important in Chapter 4 when we examine semiotics and reference in the narrative articulation of scientific knowledge.

But here what is most important is the methodological articulation of cognition within an experimental, *empirical* framework of assumptions in terms of the *narrated stages* of simple and complex binary oppositions. The attempt to trace the functioning of cognition by observing and measuring its operations as accurately as possible produces what Greimas calls a little narrative drama—"truth" in conflict with "falseness," confronting undifferentiated "nonsense" and finally saved from these dizzying complications through the agency of "meaning." The crucial question that arises from the superimposition of Greimassian semiotic frameworks upon empirical, experimental "data" is: which framework is primary? Does the logic of Greimassian understanding determine our "perception" of data (in this case conceived as "phenomena"), or does the nature of accurately measured data determine what is perceived (what constitutes phenomena) which is only afterward organized rationally?

It is impossible to decide between these alternatives, but at the same time, it is possible to shift the framework of understanding and take this very impossibility of choosing between experience and logic in viewing understanding as something to be examined. One shift, which we examine throughout this book, is to *situate* cognitive activity—to ask what particular functions it accomplishes in particular situations and what elements of unconscious power inhabits its conscious knowing as an activity that is instituted within culture. Another shift skirts these questions altogether by reducing them to irrelevance within the binary opposition between method and content. If "frameworks" of understanding—whether they focus on data or phenomena—are simply methodological ways of ordering understanding, then the question of the nature of knowledge, its components, its sources, its development, its deployment, seems irrelevant.

Binarity and History

In their definition of "binarity," Greimas and Courtés are careful to emphasize the methodological rather than an epistemological understanding of binary oppositions. Specifically, they distinguish methodological binarity from the "binarism" of Roman Jakobson, which, they write, "is an epistemological postulate according to which the binary articulation or grasp of phenomena is one of the characteristics of the human mind" (1982: 25). Central to this argument is the understanding that even the most complex structures can be "formally represented in the guise of hierarchy of binary structures" (1982: 25) and that such a method of the binary hierarchical understanding of cognitive-semiotic phenomena is, in fact, simply a historical accident.

Here we explore the conditions of the "accident" of binary conceptions of the functioning of language. After all, such methodology did not always govern understanding. Even if, as Jakobson believes, binarity is a self-evident "truth" about experience derived from universal and fundamental modes of human perception; or even if the "methodology" of binary oppositions, as in neuroscience, seems confirmed by the functioning of physical objects—the fact is striking that such epistemology and methodology arose as "privileged" in the twentieth century. The conception of binary oppositions that Greimas and Courtés describe is most evident in the synchronic analysis of language Ferdinand de Saussure pursued at the beginning of the twentieth century. Saussure articulated binarity in opposition to nineteenth-century diachronic linguistics that satisfied itself with simply recording the accidents of history. Diachronic linguistics, like Newtonian physics and Lockean psychology and even the absolutist political order of latter-day royalty in Europe, does not question the nature of phenomena beyond the explanatory power of causal origins. Something is understood if we can merely understand the forces that originate it. This is the great and powerful achievement of nineteenth-century Western intellectual culture: its ability to discover and articulate the origins of a host of cultural phenomena. In this regard Hugh Kenner has argued that the great epic of the nineteenth century is the *Oxford English Dictionary* with its discovery of the history of particular and individual words. Kenner contrasts this document to the epic of the eighteenth century, Gibbon's *Decline and Fall of the Roman Empire* (articulating as it does the Enlightenment ideology of universal human nature), and the primary epic of the early twentieth century, Joyce's *Ulysses,* with its combination of almost incomprehensible phenomena and its intricate structure of binary relationships between Greece and Ireland, Mozart and Homer, young man and older man, man and woman, and so on (1978: 49).

Such binary structures lend themselves to formalism in literary studies as well as in cognitive and other science. In fact, the formal conception of language and discourse Saussure articulated has governed the study of cultural phenomena outside the sciences since the advent of modernism in the early twentieth century. Such formalism creates the vocabulary for a significant reorientation across cultural phenomena concerning the nature of knowledge itself. Such is Bruno Latour's claim about science in general as a conclusion to his study of Einstein's relativity—and what we could say about cubism in art or, more abstractly, the privileged hegemony of binary structures in articulating understanding: "we refuse meaning to any description that does not portray the *work* of setting up laboratories, inscription devices, networks" (1988: 26). In this way, we can reasonably argue that early formalisms—Russian Formalism in the 1920s, the New Criticism in the 1930s, even Prague structuralism and the Paris semiotics of Greimas and his colleagues, and the host of disciplines constituting cognitive science—are themselves part of a larger cultural phenomenon, that of modernism, and that as such the "historical and pragmatic factors" governing the formal binarity of these approaches to discourse can be understood beyond the terms of intellectual history Greimas and Courtés describe.

Contemporary understandings of cognition, we are suggesting here and will demonstrate in much of what follows in this book, take their places among the formal renderings of experience that are crucial to cultural modernism in Europe and the United States. Modernism, as has been argued elsewhere (see Schleifer 1990: chap. 2 and 1991), is a Western response to the perceived breakup of European civilization in the twentieth century. Such a "breakup" can be seen in reorientations in intellectual history, specifically in the transformation of causal modes of explanation into the teleology of synchronic linguistics, quantum physics, Nietzschean and Heideggerian metaphysics, psychoanalysis, and the various branches of cognitive science. Jakobson's coining of the term "structuralism" in 1929 takes its place within this configuration of intellectual phenomena, as does Turing's demonstration in 1936 of the possibility of modeling complex calculations in terms of binary operations.

To give a single example of the scope of these reorientations, we can look at the conceptions of "space" in the early modern era. At the turn of the century, the complementarity of space and time broke down. (This is similar to the breakdown of the complementarity of truth and meaning we examined earlier.) Space came to be in the sciences as well as the arts, not simply a void but "a constituent function" so that there was "no clear distinction between the plenum of matter and the void of space" (Kern 1983: 177, 183). In terms of what Stephen Kern calls "positive negative space," space itself became a *material constituent* and "had one feature in common with the progress of political democracy, the breakdown of aristocratic privilege, and the secularization of life at this time: they all leveled hierarchies" (1983: 153, 177). One effect of this new conception of space, Kern argues,

> was a leveling of former distinctions between what was thought to be primary and secondary in the experience of space. It can be seen as a breakdown of absolute distinctions between the plenum of matter and the void of space in physics, between subject and background in painting, between figure and ground in perception, between the sacred and the profane space of religion. Although the nature of these changes differed in each case, this striking thematic similarity among them suggests that they add up to a transformation of the metaphysical foundations of life and thought. (1983: 153)

Such a transformation, most profoundly, made binarity noticeable: the breakdown of "absolute distinctions"—such as those between mind and reality, between meaning and truth, between science and the *work* of science, between the rulers and the ruled—allowed the binary oppositions that governed these distinctions to appear.

The occasions for this "modernist" reorientation go beyond such intellectual formations to include the transformation from industrial capitalism to finance capitalism in the years between 1880 and 1914 and the concurrent explosion of a bewildering array of absolutely new goods, and they also include the rise and fall of European imperialism, and the great and overwhelming wars—what George Steiner calls "the 'Thirty Years' War' from 1914 to 1945" (1971: 29)—of our time. This configuration of social and intellectual forces both helped condition and took its place within a sense of modernism that conceives of itself as faced with its Other—whether it be colonial peoples, women, workers, and Jews in Europe itself, the "void" of space, or, more generally, a "dehumanized" Other to the ideology of progress that seemed both the cause and effect of the liberal humanism that governed the rise of capitalism and European power from Waterloo to Sarajevo. Modernism responded to the emergence of the binary opposite of the world it inherited in that it responded to a world in which the stable verities of the past, outside the seeming struggle of binary opposites, were destroyed.

In our own era, as Nietzsche says in *The Use and Abuse of History* (narrating binarity in metaphysical terms), the knowledge of "'being' is merely a continual 'has been,' a thing that lives by denying and destroying and contradicting itself" (1957: 6). Formalism in literary and cultural criticism participates in modernism by attempting (as in T. S. Eliot's suggestion of a "mythical method" in modernism) to recuperate meaning and significance in the face of the confrontation of the Other, in the face of the "end" of European world hegemony in politics and economics and culture, what Oswald Spengler called the decline of the West. Modernism attempts to salvage something in the chaos and futility of the modern world—"O Lord," Yeats writes in *A Vision,* "let something remain!"—and most often the modernist effort at salvaging "culture" takes the form of the "work" and assertion of a hierarchical order, the assumption of "the guise of hierarchy of binary structures" in what Christopher Norris aptly calls the "aesthetic ideology" everywhere present in Anglo-American modernism (1988). Against the rage for order inherent in such aestheticism one should recall Paul de Man's striking reading of Nietzsche: "'Only as an *aesthetic phenomenon* is existence and world forever *justified*': the famous quotation, twice repeated in *The Birth of Tragedy,* should not be taken too serenely, for it is an indictment of existence rather than a panegyric of art" (1979: 93).

The "historical and pragmatic" reasons for the ascendant power of the formalism of a binary methodology responds to the conception and, more than this, the experience of life and being as a vast and bewildering disorder. Mikhail Bakhtin, from his vantage point as a historical-materialist, describes the "alarming instability and uncertainty of [the] ideological word" in the modern world, what he calls "the stage of *depression in the thematic value of the word*" (1986a: 159). It is precisely the felt absence of a European "center" of human values and culture in the modern (and postmodern) era, a "center" which had traditionally anchored and justified the accidents of history. In this "tradition" linguistics could study the development of language without worrying about the value of its enterprise, and epistemology could talk about the subject of knowledge without questioning the nature and self-identity of that subject. The lack of a center governs the depression of value Bakhtin describes. More than Bakhtin, de Man emphasizes the futility and meaninglessness of the modern world when he emphasizes (1986: 14) how grammatical and rhetorical understandings simply cannot be ordered hierarchically, one made more basic than the other. Such a vision of bewildering plenitude—for this is what de Man means when he describes the "undecidability" between grammatical and rhetorical understandings—can even be discerned, as Fredric Jameson argues, in the methodological rigors of Greimas's semiotic square, which, he says, "opens up a dizzying perspective of the subatomic universes, a prospect of what a very different semiotician, Umberto Eco, following Peirce, calls 'infinite semiosis'" (1987: xvi). In fact, it is just such a vertiginous vision of infinite semiosis—in finance, in labor, in cognition, in the whole order which Jean Baudrillard describes as "the reigning scheme" of our era, the scheme of "simulation" (1983: 83)—that binarity resists.

Binarity and Truth

The privileged place of structural accounts of cognition, governed as they are by binary oppositions (unlike Bakhtin's accounts, which are about "the extreme *heterogeneity* of speech genres" [1986b: 60]), may indeed be pragmatic and accidental, but that "place" itself exists in a historically determined context. Binary structures—those of one-to-one opposition—are a function of the perception of the breakdown of hegemonic self-evident truths. When truths are self-evident, there is no need self-consciously to define the true against its opposite. (There is also no need to distinguish between truth and meaning or epistemology and method.) As Derrida has attempted to show repeatedly, binary oppositions are "never the face-to-face opposition of two terms, but a hierarchy and an order of subordination" (1982: 329). De Man also notes that "binaries, to the extent that they allow and invite synthesis, are therefore the most misleading of differential structures" (1986: 109). In this way, the historical and pragmatic factors that have given binary structures a privileged place in linguistic methodology arise in relation to the breakdown of the hierarchic ordering of experience and perception at the end of the nineteenth century and the felt need, in the face of what Eliot describes as this "futility and anarchy" (1975: 177), for reestablishing hierarchical order. Binaries are most misleading because they present, in all their "oppositions," a sense of truth and countertruth at play, while, at the same time, they covertly reestablish hierarchical power. They create the sense that one of the elements of its binary opposition can always be seen to embody the general term that encompasses the *whole* of the opposition (see Schleifer 1987b).

The greatest proponent of the functioning of binary structures in discourse and semiotics, as Greimas and Courtés suggest, is Roman Jakobson. Unlike other twentieth-century semioticians, he consistently argued for the isomorphic functioning of binarism across *all* levels of the functioning of language, from the "distinctive features" that, in "bundles" of binary oppositions, comprise the phonemes of language, to the binary oppositions that govern the semantics of poetry and discourse in general. For this reason, Lévi-Strauss was able to develop the narratology of structuralism based on Jakobson's binarism, and Greimas was able to develop his elementary structure of signification by working out Lévi-Strauss's critique of Vladimer Propp in "Structure and Form" in his analysis of Propp in *Structural Semantics*. (As Greimas later noted, Propp's *Morphology of the Folktale* is a special case of a more general semionarrative understanding of discourse.) Thus, there is something ingenuous in Greimas's recent claim to be a "poststructuralist" (1989: 539). In fact, he could easily join Jakobson and Saussure as the objects of Derrida's early deconstructive critique of the "scientificity" of linguistics in *Of Grammatology* (1976: 29). Even so, in an important moment of his early critique of Hegel, Derrida presents what can be taken as a narrative description of the semiotic square in his assertion that "the *point of nonreserve* . . . cannot be inscribed in discourse, except by crossing out predicates or by practicing a contradictory superimpression that then exceeds the logic of philosophy" (1978: 259).

The power of such formal structures, not only in Jakobson and Greimas, but in the understanding of cognition itself, is the rhetorical power of the spatial figures that create the meaning-effects of "knowledge"—the self-evident "recognitions" of cognition—by conceiving of phenomena as "substantial." This conception makes the objects of cognition, as Latour says, both permanent and transportable from context to context. In *The Use and Abuse of History* and elsewhere, Nietzsche attempts to show such "knowledge" to be "willfully" produced. Greimas narrates this process of producing meaning by describing the tendency of discourse to "substantify" relationships so that "whenever one opens one's mouth to speak of relationships, they transform themselves, as if by magic, into substantives" (1970: 8; see also Schleifer 1987a: 40–43). Greimas is describing the nature of language which, as he sees it, creates the effect of cognition by giving rise to phenomenally "substantial" referents for its designations. When meaning is apprehended, even when it articulates a relationship, the act of designating that relationship causes it to seem less like a method of "producing" meaning and more like a simple perceiving of preexisting "data." For instance, Saussure's word *signifier* (i.e., the French *signifier,* literally the "signifying") transforms relationship into a spatially locatable substance. Or, in another instance, methodology—such as the methodology of deploying binarity Greimas and Courtés describe—transforms itself, as if by magic, into epistemology.

The power of substantification (Latour's "inscription") answers flux, becoming, and overwhelming uncertainty—in larger cultural terms, it responds to the breakdown of European world hegemony—with the rhetorical effect of "objective" scientific phenomena which can be cognitively manipulated and configured. This is why, as Bakhtin says, "the *declaratory* word remains alive only in scientific writing" (1986a: 159). In such writing the existence of things remains "alive" precisely because, in the social network of cognitive "truth," they transcend, like Eliot's "tradition," the temporalities of narrative discourse. This is also the reason that Anglo-American cognitive science makes so much of hierarchical levels. Greimas also emphasizes hierarchical levels, going so far as to assume the binary oppositions of the rationalism of Chomskyan linguistics—"deep level" vs. "surface level" (Greimas and Courtés 1982: 134); "competence" vs. "performance"; the "generative" nature of semiotics (Greimas 1989a: 540)—even though Chomskyan rationalism (Cartesian rationalism that substantifies "mind") is far from the tradition of Continental semiotics within which Greimas and his colleagues work.

Greimas is an instructive example of the relationship between cognition and narrative precisely because he brings many of the assumptions of cognitive science to the study of narrative structures themselves. Beginning in *Structural Semantics*, Greimas explicitly analyzes discourse in terms of its articulated or "substantified" *agents*—what he calls the "actants" of narrative—rather than its represented *activities*—what he calls (following Propp) the "functions" of narrative. By focusing on a rhetoric of space and inscription, Greimas demonstrates a greater possibility of analysis and, of course, of cognitive configuration (1983b: xxxviii). In a word, without some kind of cognitive "mediating structures," conceived more or less consciously in spatial metaphors, the "social system" of language, as Greimas and Courtés note in *Semiotics and Language* under the heading "Enunciation," can only be "scattered into an infinite number of examples of speech (Saussure's *parole*), outside all scientific cognizance" (1982: 103). It is precisely such scattering that was the perceived experience of European modernism.

Narration and Cognition

Like Greimas, many others have pursued the certitude of binary opposites in response to this experience of modernity. We have seen the similar oppositions—between truth and meaning, between meaning and unmarked nonsense—in the attempts of cognitive psychology to measure and describe the functioning of metaphors. For this reason, Derek Attridge situates Jakobson's poetics within an opposition between conceiving of poetry as functioning "to heighten attention to the meanings of words and sentences" and conceiving of poetry as "a linguistic practice that specially emphasizes the material properties of language . . . independently of cognitive content" (1988: 130). This opposition, which Attridge suggests functions unconsciously in Jakobson, puts a "formal" and "cognitive" understanding of meaning against a "material" and "phenomenal" understanding. In doing so, it suggests that even the seemingly phenomenal effects of language can be accounted for within the structures of binary opposition and that, in Derrida's paraphrase of Edmund Husserl, "meaning is everywhere" (1981: 30). That is, in doing so, it asserts (in the face of its binary opposite) that the real, as Whitehead describes it, is rational (Schleifer 1990: 28).

The same opposition can be seen in Anglo-American cognitive psychology more generally. In the last three decades, in response to the hegemony of behaviorism in psychology, cognitive psychology, as we have seen, has reintegrated cognition and meaning into the science of psychology in the same way that Jakobson's project and that of Prague linguistics described the basic linguistic function at all levels as "the distinction of meanings" (Jakobson 1962: 1). Bartlett argued early in the debate that memory was constructive and could be explained as "effort after meaning" (1932). Jean Piaget suggested that all life is involved in genetic epistemology which actively constructs systems for acquiring knowledge (see 1970: 52–73). Ulric Neisser situated the subject in a social and human context to examine how we get by, how we use these systems to construct a workable reality (1982; see also 1987). A chief proponent of this reconfiguration of scientific psychology is Jerome Bruner, who, in a career that has repeatedly attempted to define meaning and understanding, has created the groundwork for studies such as that of the functioning of metaphoric cognition we examined earlier within his discipline. In *Actual Minds, Possible Worlds,* Bruner presents the same kind of opposition between cognitive and noncognitive conceptions of understanding as we find in Jakobson and Greimas. But in this work, unlike the studies of metaphor we have discussed, he examines the comprehension of meaning in relation to an opposition between apprehension and narrative, what he calls "two natural kinds" of understanding.

These "two modes of thought," these "distinctive ways of ordering experience, of constructing reality," are "a good story and a well-formed argument," what Bruner calls "the narrative mode" and "the paradigmatic or logico-scientific" mode (1986: 11, 12). The second of these "yields accounts of experience that are replicable, interpersonally amenable to calibration and easy correction" (1986: 110). The first "leads to conclusions not about certainties in an aboriginal world, but about the varying perspectives that can be constructed to make experience comprehensible" (1986: 37). Here Bruner is pursuing the old controversy between the arts and sciences, the "two cultures," and at its best his argument is able to bring these "fundamentally different" modes of verification (1986: 11) together in a coherent argument that convinces us of both its "truth" and its "lifelikeness" (versions of the "truth" and "meaningfulness" of the Gregory/Mergler study). Of Jakobson's contention that the nature of "literariness" is to make "the world strange again," Bruner argues that "this ingenious intuition can be given a psychological rendering [open to] . . . empirical research" (1986: 75). Throughout *Actual Minds, Possible Worlds,* Bruner offers empirical studies to substantiate narrated theories. He uses the linguistics-based theories of Tzvetan Todorov—theories governed by the structuration of binary oppositions—to distinguish between narrative and expositional discourse, and then discusses an experiment in which subjects renarrated Joyce's "Clay" to describe the narrative mode of thinking. In another instance, he returns to one of his own early experiments, which examined how subjects construct abstract "realities" or "wholes" from amorphous experience, and discovers, on retesting, what he had overlooked in the 1950s: that subjects can use either of the two modes of cognitive processing to produce "world constructions" (1986: 89–92).

Most interesting to us here is the fact that Bruner, like Jakobson and de Man, shapes his understanding by means of binary oppositions—the means of the paradigmatic or logicoscientific mode of understanding—even while he does so in the context of a narrative account of his career and the career of cognitive psychology more generally. Thus, in the most impressive chapter of the book he narrates "how the three modern titans of developmental theory—Freud, Piaget, and Vygotsky—may be constituting the realities of growth in our culture rather than merely describing them" (1986: 136) in order to narrate the *situation* of cognitive psychology in a world subject to nuclear destruction, a world that needs to construct the narrative of its future (1986: 149). That is, here, as in Jakobson and de Man, there is a constant pressure of binary oppositions to mark themselves as positive and negative and, as we mentioned earlier, covertly reestablish hierarchical power. To do so, they articulate the cognitive understanding of binarity—even when, as in moments in Bruner, the "good story" of "the narrative mode" seems to take precedence.

Jameson, as we have noted, emphasizes a similar opposition in Greimassian semiotics when he describes the dialectic between the "profound narrativity of all thinking" in Greimas's work—an aspect of his work that focuses on the *production* of meaning—and the "specialized abstract" cognitive mode of thinking that equally governs his understanding of meaning. This dialectic is clear in Greimas's recent work which explores modal understandings of meaning, and which, as Jameson notes, assumes that mental processes are cognitive. At crucial moments, even studies most fully governed by the methodology of binarity transcode the level of analysis to that of narrative. In Greimas's "lexical semantic" study of anger, for instance, after a detailed semantic analysis that repeatedly produces more and more minute binary discriminations (see the explosion of binary oppositions in a single paragraph of the analysis [1987: 150]), the analysis suddenly shifts to terms of narrative analysis. Moreover, in a short section of this study Greimas narrates the very act of cognition by describing three different levels whose order of presentation offers a narrative progression that recapitulates the narrative trajectory of his semiotics. Here his analysis is close to the narration of three levels of metaphoric cognition in the Gregory/Mergler experiments. First, Greimas presents an "empirical" or "accidental" reading of anger in an analysis based on the narrative functions (abstract invariable narrative "events") developed by Vladimir Propp (1987: 161). Then he focuses on the meaningfulness of this narrative by investing Propp's function of a "glorifying test" with a new semantic meaning. He does this by renaming the function of "recognition" the "cognitive sanction" of language, the reinstitution in the narrative of "the language of truth" (1987: 162). Finally, examining the terminal functions of the narrative progression of anger (the first and last Proppian functions that occur in the *activity* of anger), he opposes the "cognitive sanction" at the end of an episode of anger to an initial "fiduciary lack" at the beginning (1987: 162). In his seemingly abstract argument, he reinscribes the binary opposition between nonsense and meaning within the narrative discourse of his short analysis by replacing "causal description" with the "semantic description" of narrative (1987: 164).

In this discussion Greimas uses hierarchical binary structures—the very structures of Continental structuralism—which are quite similar to structures that inhabit Anglo-American cognitive science. Moreover, his discussion itself is an example of the way such binary structures encompass the dialectic of modernism that both articulates and resists the futility and anarchy of its historical moment. As in Bruner, the binary oppositions of Greimas's "actantial" analysis do not preclude "functional" analysis but rather attempt to create a metalanguage that allows for the complex analysis of meaning, which is always both cognitive and narrative, always both an achieved order and a struggle for order. Such complexity can be seen in modernist formalism from Eliot's attempt to recuperate a "deep" and "transcendental" meaning from the flux of Joyce's experience to the hypostatizations of the formal opposition of stimulus and response within the chaos of mindless behavior in behaviorism. Behaviorism also attempts the recuperation of transcendental value in the face of what Wallace Stevens calls "a great disorder." Continental structuralism, however, as Lévi-Strauss has argued, aims at *situating* such formalism. This aim is unlike the aims of Russian or New Critical formalisms or the behavioral formalisms of Leonard Bloomfield in linguistics and in the behavioral psychology of the 1930s more generally, all of whose forms are abstract paradigms completely separate from the actual, temporally situated phenomena they study. In all these disciplines method and truth are readily separated.

Lévi-Strauss argues, however, that structuralism arrests and apprehends the "logical organization" of phenomena "conceived as a property of the real" (1984: 167; see Galan 1985: 35 for Jan Mukařovský's earlier criticism of formalism in similar terms). That is, the structures of structuralism—including the methodological binarities of Greimassian semiotics and Brunerian cognitive science—do not simply assert the existence of nonmaterial abstract entities such as Northrop Frye's opposition between comedy and tragedy figured as the opposition between Spring and Fall; rather, they attempt to recuperate, modally and functionally, the phenomenological and temporal processes of apprehension and organization. Like the opposition of presence and absence in the distinctive features of phonemes, they allow the phenomena studied to exist *as* phenomena both in time and in cognitive apprehension. In this, they allow for logical structures and a plurality of cultures. The relationship between cognition and narration is, as Greimas says of narrative in general, "neither pure contiguity nor a logical implication" (1983b: 244). For this reason, phenomenal meaning-effects exist and function as narrative units *because* logical organizations are apprehended as real temporal events—cultural events. Here the binary opposition between narrative and cognition, method and epistemology, story and argument, breaks down precisely because this binary opposition itself is both true and false—which is to say, both true and meaningful, perception and heurism—at the same time.

The felt sense of opposition within any articulation of meaning—the method (which is also more than a method) of binarity—is central to semiotics and cognitive science in the same way it is central to the modernism and postmodernism of our particular time. This opposition exists, as does Greimassian semiotics and Brunerian cognitive psychology, because at this historical moment the sureties of order and the self-evidence of value are in question. The work of semiotics and cognitive science—the work of psychology itself in its broadest definition, which encompasses the perception of truth, the generation of meaning, and the problematics of subjectivity—articulates this situation. It *enunciates* it (even in its "scientific" language and often magisterial pronouncements) in its constant oscillation between transcendental cognition and situated narrative. Semiotics schematizes this opposition, substantifies it, and narrates it within the processes of analysis and understanding. In the next two chapters we examine particular structures conditioning such enunciations in Greimassian semiotics and in empirical accounts of the cognition and discourse of old people. These chapters attempt to describe the *simplicity* of the logic of discourse as articulated in Continental semiotics and the *empirical* functioning of cognition in a survey of data concerning cognition at a particular moment in human development. They take their places in our larger attempt to articulate the relationship between cognition and narrative—the connection between general and special cases of understanding—and to describe cognitive institutions of understanding in our time.

2

STRUCTURES OF MEANING

The Logic of Narrative and the Constitution of Literary Genres

Cognitive Semantics

In this chapter we examine a logical Greimassian model that attempts to describe the phenomenon of discursive cognition, what Paul Ricoeur calls the attempt "to superimpose a logical type of rationality on the intelligibility that already lies in the production of narratives" (1985: 29). This is a simplifying model in which the seeming heterogeneity of literary narratives is reduced to a six-element schema, yet insofar as it deals with the complexities of narrative, it rigorously complicates the "simplicity" of logical binarity in its account of semantic cognition. In the next chapter we survey the results of experimentally controlled empirical situations testing the cognitive skills of old people to examine cognitive behavior in a way that complicates the "simplicity" of empiricism in its account of social and cultural semantics. The model we examine here is based on the cultural anthropology of Claude Lévi-Strauss; that of the next chapter is based on the diachronic taxonomies of Charles Darwin. In these chapters we are presenting the logical rationalism of Continental semiotics and the empirical functionalism of Anglo-American cognitive psychology by focusing on cognition in relation to narration. These two models—Continental rationalism and Anglo-American empiricism—attempt to define and understand cognition from two different vantages. In Part II and Part III of the book we retraverse this terrain. The rational formalism of genres we describe in the present chapter makes clear the cognitive implications of rationalism by defining what we are calling Greimas's cognitive semantics within the complications of narrative. Greimas describes such narrative, as we have seen, as "neither pure contiguity nor a logical implication" (1983b: 244): neither the simplicity of chance nor the simplicity of logic. In the next chapter we similarly *complicate* simple empirical accounts of cognition by situating cognitive activity within interhuman relationships.

Still, Greimas, unlike the cognitive scientists, defines cognitive activity rather formally, in what he calls the abstract "actants" of narrative discourse. These actants are defined, more or less, in terms of "logical implication." Yet Greimas is able to situate their meaning within the complications of the semiotic square, and by so doing he is able to include within his analysis "external" (or "contiguous") cultural criteria affecting their signification. That is, he includes the actants of "sender" and "receiver"—agents of discursive activity—within the logic of his semantics, and by so doing he superimposes logic and contiguity, meaning and truth. Most specifically, he suggests that narrative genres can be defined in terms of the "receiver" of cultural meaning, cultural power, cultural goods, just as in the next chapter we explore the possibility of defining the function—the "logic" of the actantial role—of old age within the sprawling contiguities of human communities in terms of the "sender" of cultural meaning, cultural power, and cultural goods. Unlike Louis Hjelmslev, who attempts to articulate the logical relationships that obtain between the elements of language while bracketing the "content-substance" of language's referents (1961: 79; see Sampson 1980: 167), Greimas's semantics attempts to inscribe "content-substance" within language. Such an aim is the *cognitive* project of genres altogether, which attempt to articulate and classify thematic-formal aspects of literary discourse. That is, genre categories attempt what semiotics and cognitive science attempt, to articulate structures of signification, of apprehended meaning. Thus, the controversies and arguments that surround generic studies reproduce those that surround semantics and cognitive science in general. Not only Greimas, but Lévi-Strauss (along with Freud, Lacan, Bruner, and even Northrop Frye and Alastair Fowler), can be said, in these terms, to be cognitive "semanticists." Greimas's semantics attempts to go not only beyond the confines of the sentence but beyond the confines of an abstract formalism such as Hjelmslev's and a concrete formalism such as Frye's. Actantial analysis of narrative, he argues, describes both the qualities and the activities of actants. Describing qualities, it describes particular contents and delineates modes in terms of relationships between and among the actants. Describing activities—he calls them "functions"—it describes narrative as such and delineates narrative genres in terms of the "action" of sender and receiver and the cultural values—meaning, power, goods—they traffic in.

The Logic of Discourse

The work of Claude Lévi-Strauss is an important contribution to the study and science of cognition, especially in relation to narrative and culture. His study of narrative forms across very different cultures is an attempt to discover the cognitive structures that govern the understanding of narrative discourse. As Edmund Leach has noted, the "general object of analysis" in Lévi-Strauss's work, like that of the semiotics and cognitive science we examined in the preceding chapter, "is conceived as a kind of algebraic matrix of possible permutations and combinations located in the unconscious 'human mind.'" Lévi-Strauss, Leach goes on, "conceives of the 'human mind' as having objective existence; it is an attribute of human brains. We can ascertain attributes of this human mind by investigating and comparing its cultural products" (1970: 40). For this reason, he presents a Boolean matrix, "a structural model defined as the group of transformations of a small number of elements," in what is perhaps his most condensed articulation of his project of narrative and anthropological structuralism, "Structure and Form: Reflections on a Work by Vladimir Propp" (1984: 183). This essay appeared in 1960 soon after the English translation of Propp's *Morphology of the Folktale* appeared.

The Boolean matrix Lévi-Strauss presents in "Structure and Form" (Figure 1) is based on the work of the nineteenth-century mathematician George Boole. In *The Laws of Thought* Boole attempted to describe the basic laws of thought in terms of the principles of logic. To this end, he used a set of arbitrary symbols (*W, X, Y, Z,* and so forth). Such an attempt, he wrote, would "express logical propositions by symbols, the laws of whose combinations should be founded upon the laws of the mental processes which they represent"; these symbols, he concluded, would "be a step toward the philosophical language" (cited in Gardner 1985: 143). The procedures of Boole's analysis, Gardner notes,

> amounted to a kind of "mental algebra," where reasoning could be carried out in abstract positive or negative terms, unsullied by the particular associations tied to specific contents. . . . Most important for the future, Boole

> observed that his logic was a two-valued or true-false system. Any logical expression, no matter how complex, could be expressed either as 1 (standing for "all," or "true"), or as 0 (standing for "nothing," or "false"). The idea that all human reason could be reduced to a series of yes or no discussions was to prove central for the philosophy and science of the twentieth century. (1985: 143)

As we have seen, Greimas's semiotic square uses and complicates the binarity that Gardner is describing. Equally important, the work of Boole influenced Anglo-American philosophy and science far more than it did the Continental tradition. Gardner notes how heavily Whitehead and Russell relied on the formalism pioneered by Boole in their *Principia Mathematica*, citing Russell's comment that "pure mathematics was discovered by Boole in a work he called 'The Laws of Thought'" (1985: 143). That is, Boole's formalism is the site of the intersection of important work in the two traditions we are examining.

Lévi-Strauss, as we noted, develops a Boolean matrix in his critique of Propp's more or less empirical account of wondertales in *Morphology of the Folktale*. In the *Morphology* Propp examined one hundred Russian wondertales and postulated that the various activities of the characters of these tales could be reduced to thirty-one narrative "functions." These functions were abstract descriptions of plot actions: for instance, in many stories a "magical agent" is given to the hero of the story so that one narrative function is defined in the statement "the hero acquires the use of a magical agent" (function 14; Propp 1968: 43), even though in different stories the agent itself is different (a magic wand, a secret ring, etc.) and the character who gives the agent is different (a witch, a bird, even chance). Most remarkable in Propp's analysis is that whenever a function was present in the Russian wondertales, it would occur in the same sequence in the plot: if the thirty-one functions of narrative were numbered, a higher number never preceded a lower number in the plot of different narratives. Besides the functions, Propp isolated seven "dramatis personae" of the wondertales, that is, seven roles or what he called "spheres of action" (1968: 79) that defined the particular characters inhabiting different stories. In doing this, Propp used the functions to define the personae: the roles were "spheres" of functional activity.

The appearance in English of Propp's 1927 Russian study was an important event for Lévi-Strauss because it seemed to confirm his own sense that the overwhelming variety of narratives found in all cultures throughout the world could be reduced to a structural model in much the same way the great variety of sentences in any natural language can be reduced to the structural model of grammar. Propp had, in fact, devised methods of narrative analysis very close to the methods of linguistic analysis that were developed by the Prague School of Linguistics a decade later (see Liberman 1984; and Schleifer 1987a). Lévi-Strauss himself had been greatly influenced by the work of Roman Jakobson, and the appearance of *Morphology of the Folktale* gave Lévi-Strauss the opportunity, as we have said, to articulate his program of structural anthropology.

In "Structure and Form," Lévi-Strauss discusses the limitations of the formula Propp had propounded in *Morphology*. Specifically, in place of the fixed order of the thirty-one functions Propp had isolated in the Russian wondertale, Lévi-Strauss substitutes the Boolean matrix shown in Figure 1 (1984: 183).

$$\begin{array}{lllll} W & -X & 1/Y & 1-Z & \ldots\ldots \\ -W & 1/X & 1-Y & Z & \ldots\ldots \\ 1/W & 1-X & Y & -Z & \ldots\ldots \\ 1-W & X & -Y & 1/Z & \ldots\ldots \end{array}$$

Figure 1

Although Lévi-Strauss simply offers the matrix as a "structural model," its elaboration clarifies the concept of "structure" in terms of *positions*. What the matrix does—what the structures of structuralism do—is to create empty slots or positions which are defined, diacritically, in relation to other positions in a structure. In this matrix different symbols—W, X, Y, Z, representing different narrative functions—are inscribed in a network which also defines a system of logical relationships—W, $-W$, $1/W$, $1 - W$—among those events, a system of their possible transformations. Thus, the matrix offers a paradigmatic network of logical relationships among narrative events—a logic of discourse in W, $-X$, $1/Y$, $1 - Z$ rather than the fixed chronological (syntagmatic) order Propp described and numbered in the *Morphology*.

Greimas has defined the network of logical relationships, of structural positions, as the "elementary structure of signification." He calls this structure "a double relation of *disjunction* and *conjunction*" and, as we have seen, he inscribes it in the "semiotic square" shown in Figure 2 (1987: 49).

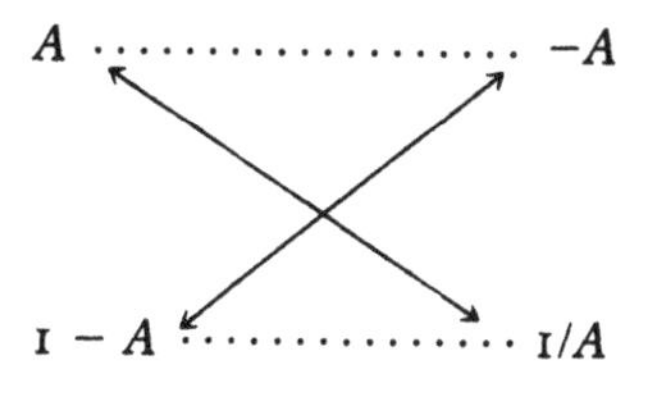

Figure 2

The schema inscribes three logical relationships: *contrary* relationships of the reciprocal presupposition of binary opposition ($A \approx -A$); *contradictory* relationships of the "absolute absence" of the elements of the binary opposition ($A \approx 1/A$) in which the latter position ($1/A$) is categorically different from the former; and finally *complementary* relationships of simple presupposition or implication ($A \approx 1 - A$ and $-A \approx 1/A$) in which the latter positions ($1 - A$ and $1/A$) presuppose the existence of the former (A and $-A$) (1987: 50–52). These three logical relations—which, in the context of phonology N. S. Trubetzkoy has denominated "privative," "equipollent," and "gradual" or "arbitrary" (1969: chap. 3)—exhaust the logical possibilities of binary opposition. A *contrary* (or "privative") relationship creates a double relation of conjunction and disjunction in terms of the presence or absence of some shared *feature;* a *contradictory* (or "equipollent") relationship creates that double relation in terms of a shared *function;* and a *complementary* (or "gradual") relationship creates that double relation in terms of a shared *situation* (that is arbitrarily defined). Thus, in phonology, as Geoffrey Sampson notes, Trubetzkoy

> distinguishes between (i) *privative* oppositions, in which two phonemes are identical except that one contains a phonetic 'mark' which the other lacks (e.g. /f/ ≈ /v/, the 'mark' in this case being voice), (ii) *gradual* oppositions in which the members differ in possessing different degrees of some gradient property (e.g. /I/ ≈ /e/ ≈ /ae/, with respect to the property of vowel aperture), and (iii) *equipollent* oppositions, in which each member has a distinguishing mark lacking in the others (e.g. /p/ ≈ /t/ ≈ /k/). (1980: 108)

The Boolean matrix Lévi-Strauss presents in "Structure and Form" elaborates these relationships to suggest a logic of narrative discourse.

Following the basic hint of a logic of discourse Lévi-Strauss presents in this essay, Greimas takes great pains in *Structural Semantics* to reduce Propp's thirty-one chronological narrative functions to the three logical categories. (In *Structural Semantics* he first formulates the relationship that became the "semiotic square" [1987: 49].) To give one example of this reduction, we can note that Greimas argues that Propp's narrative function 5, *information,* is logically contrary to *inquiry* (function 4), and it is contradictory to *recognition* (function 27), whose own contrary, *marking* (function 17), implies *information.* (In *Structural Semantics* Greimas's translations of the names of Propp's functions into French from the English translation of *Morphology* are not always precise; see 1983b: 223–24.) Greimas's description of the logical relations among Propp's functions are mapped in the semiotic square presented in Figure 3.

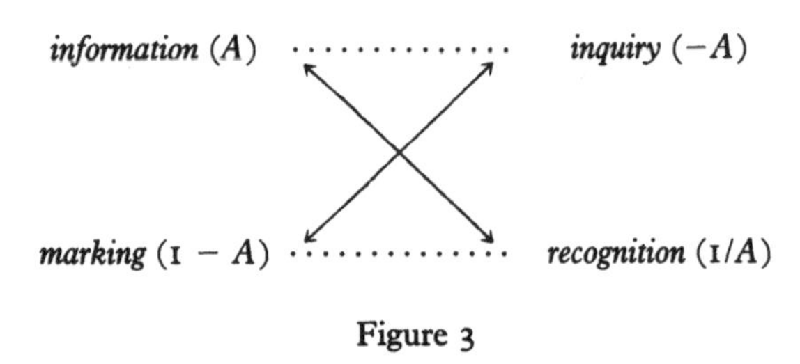

Figure 3

Information is the binary opposite to the lack of information *inquiry* implies. Moreover, it is contradictory to *recognition* in that this opposition suggests the presence and absence of constative truth (versus performative meaning). That is, *recognition,* like *marking,* semantically suggests a subject whereas *information* does not. (For function 4 Propp's English translators used *reconnaissance* which Greimas changed to *enquête* at least in part to underline the nonsubjective constative nature of this function.) The four functions, related in this way, create one moment in what Greimas calls the discursive category of "communication" or "knowledge" that transcends the necessary temporal ordering that Propp claims for his numbered functions (see 1983b: chap. 11 and xlvii).

Greimas has related four narrative functions, W, X, Y, Z, in a logic of discourse, $W, -X, 1/Y, 1 - Z$. In so doing, he creates the basis of an "actantial" analysis of narrative, what might be called a "semantic syntax" of narrative. Actantial analysis, however, delineates more than a grammar for the syntax of narrative events. What Greimas defines as "actants" are constant narrative roles that recur in the discourse of narrative events and functions—roles which particular narrative actors assume in narrative in the same way that particular words and phrases in sentences assume syntactic roles (e.g., subject, object). "If it is remembered," Greimas notes, "that *functions,* according to traditional syntax, are only roles played by words . . . a proposition, in such a conception, is indeed only a drama which *homo loquens* produces for himself. The drama has, however, this peculiarity, that it is permanent: the content of the actions is forever changing, the actors vary, but the dramatic utterance stays always the same, for its permanence is guaranteed by the unique distribution of its roles" (1983b: 198). Lévi-Strauss's matrix and the logical relationships it describes can, thus, be seen to inscribe the roles assumed by the characters of narrative as well as the events performed by them. In fact, in his critique of Propp, Lévi-Strauss says that each of the characters of narrative "far from constituting a single entity—forms a bundle of distinctive features like the phoneme in Roman Jakobson's theory" (1984: 182). These distinctive features relate the characters to other characters diacritically (paradigmatically) so that the characters of Russian wondertales—and of myths as Lévi-Strauss describes them—not only do not constitute single entities but cannot exist outside the relationships among the roles of discourse.

As we have suggested, in terms of narrative roles—in terms of *actants*—Lévi-Strauss's grid can be read two ways, as sets of individual roles designated *W, X, Y,* and *Z*, and as a set of relationships between those roles designated W, $-W$, $1/W$, and $1 - W$. Such a double reading, we will argue, can achieve what Greimas calls "the development of a general theory of genres" (1987: 113; see also Greimas 1970: 249–70) by effecting the distribution of actantial roles (W, $-W$, $1/W$, $1 - W$) to the characters or actors of discourse (*W, X, Y, Z*). It can do so, moreover, because, as Greimas notes elsewhere, genres are not simply a particular "zone" of signification but rather a realization of the structure of language. "The domain of literature," he writes, "distinguishes itself from other domains (religion, law, etc.) not because it is characterized by a particular zone of content-substance. On the contrary, the content-"forms" which seem at first to define its domain (tropes and genres) are metalinguistic in relation to particular natural languages [*naturelles langues*] and form a part of the general structural properties of language in general [*langage*]" (1970: 271–72; our translation). Greimas is asserting the relationship between genre and the elemental structure of signification. Actants are situated between what Greimas calls these "deep structures" and "discursive structures" of particular texts, on what he calls the "semionarrative" level, and as such, like genre, they mediate between meaning and manifestation (1987: 88, 140; see also Schleifer 1987a: chap. 3). In this way genres can be conceived and "developed" structurally in actantial terms.

Actants

Lévi-Strauss himself did not intend to restrict the possibilities of his matrix to four elements. Its purpose in "Structure and Form," in fact, is not to designate narrative roles but narrative functions, and no structural analysis reduces the basic roles of narrative discourse to only four. Propp, as we have noted, designates seven narrative roles, which he calls the "dramatis personae" derived from their "spheres of action" in narrative: villain, donor, helper, princess or sought-for person and her father, dispatcher, hero, and false hero (1968: 79). In *Structural Semantics* Greimas reduces Propp's seven dramatis personae to six "actants," his designation of basic narrative roles. He models the actants on grammatical syntactic categories, and he articulates them within three relational categories: subject vs. object, sender vs. receiver, and helper vs. opponent.

The first category is most clearly grammatical. The subject and object of narrative are positioned analogously to the subject and object of a sentence. The second category is syntactical in the most global way: the sender and receiver of narrative are positioned analogously to the addressor and addressee of discourse, what Greimas and Courtés later describe in *Semiotics and Language* as the enunciator and enunciatee. In the final category helper and opponent are positioned like sentence modifiers. In *Structural Semantics* Greimas describes them as analogous to adverbs, although in his later work he abandons this category and replaces it with his elaborate analysis of the modalities of discourse. We will keep his early designations, both because they clearly are based on Propp's and Lévi-Strauss's narrative analyses and because they represent a special case of Greimas's later modal analysis—namely, the modality of */being-able/*—that is most pertinent to plotted narrative. Greimas claims that these six actants are found in all forms of narrative, including what he calls the "narrative program" of other forms of human relationships—a businessman's relationship to his firm, for instance, or a revolutionary's conception of his relationship to the class struggle (1983b: chap. 10; see also Greimas and Landowski 1976: 79–80).

A comparison of the "actants" of Propp and Greimas suggests that four of Greimas's actantial roles position themselves within a logical network that implies an actantial typology of literary genres (see Figure 4).

Propp	*Greimas*
Hero	subject
Princess (sought-for person <and her father>)	object
Villain False hero	opponent
Helper Donor	helper
Dispatcher <father>	sender
————	receiver

Figure 4

From this comparison it is clear that Propp and Greimas share four actantial categories: villain (false hero)/opponent, helper (donor)/helper, princess (sought-for person)/object, and hero/subject. In Greimas's actantial terms inscribed here are subject vs. object and helper vs. opponent. Propp completely leaves out Greimas's category of receiver and thereby confuses the role of the sender (which for Greimas is presupposed by the receiver) with that of the object.

This comparison is quite suggestive because the last category, sender vs. receiver, articulates the *situation* of linguistic activity, whereas the other actantial categories help describe semionarrative relationships *within* a message. "Linguistic activity," Greimas writes,

> creative of messages, appears first as the setting up of hypotactic relationships between a small number of . . . functions, actants, contexts. It is thus essentially morphemic and presents a series of messages as algorithms. However, a systematic structure—the distribution of roles to the actants—is superimposed on this hypotaxis and establishes the message as an objectivizing projection, the simulator of a world from which the sender and the receiver of a communication are excluded. (1983b: 134)

Linguistic activity begins with a relationship between sender and receiver, but the systematic structure of its message—its cognitive content—"excludes" that relationship. For this reason, it would seem to be legitimate to exclude the category sender vs. receiver from a general theory of genres determined by the systematic distribution of actantial structures.

Nevertheless, throughout his work Greimas suggests that what best characterizes literary genres are not the actants themselves—which, as we noted, he takes to be universal categories that allow the apprehension of significance beyond the limits of the sentence—but rather the "fusion" or "syncretism" of actants. Greimas notes that Propp does not "fill" the category of receiver in his description of the dramatis personae of the Russian folktale. "As for the receiver," he writes, "it seems that, in the Russian folktale, his field of activity is completely fused with that of the subject-hero. A theoretical question that can be raised about this point, one that we will return to later, is whether such fusions can be considered as pertinent criteria for the divisions of a genre into subgenres" (1983b: 204). Later, he does argue such fusions are pertinent:

> The first typological criterion . . . can well be the syncretism, often recorded, of the actants. We can thus subdivide the models into genres, according to the nature of the actants which let themselves be syncretized: in the folktale, we have seen, the subject and the receiver constitute an arche-actant, in the model of economic investment, in turn, the arche-actant is realized by the syncretism of the object and receiver, and so forth. Taken in the nonaxiological domain, the example can be made clearer: thus the queen, in the game of chess, is the syncretic arche-actant of the bishop and the rook. (1983b: 211)

While the chess example may make the syncretism Greimas is describing clearer, it also obscures the *generic* property of this phenomenon. It does so by obscuring the fact that the actantial role of receiver is not one role among others and simply exemplative; rather, it is the *particular* syncretic position of actantial analysis. Accordingly, in order to distinguish between two texts in *Maupassant,* Greimas notes that he has to make "a fundamental point, that of the choice of the *receiver*" (1988: 242).

Moreover, Greimas repeatedly notes the special status of sender as an actant and, thereby, implies the special status of the receiver. "One of the reasons for the actantial position of *Sender,*" he says in *Maupassant,* "is actually to transform an axiology, given as a system of values, into an operative syntagmatics" (1988: 44). The sender, he notes elsewhere, "properly understood, is only the incarnation at the level of anthropomorphic grammar of the universe of values" (1983a: 221). The receiver, then, is a passive role different from the others. It is "chosen" by the sender to receive what the sender never relinquishes, cultural values that are simultaneously immanent—that is, *received* by an actor—and transcendent, "inscribed within the very prescriptions of [the] language" that articulates them (Schleifer 1987a: 106). The receiver is always syncretized with another actantial role, whereas the sender both transcends the "objectivized" message and participates in it insofar as it designates one of the actants of its "systematized structure" as its receiver. Because of this, the linguistic message can never fully project "a world from which the sender and the receiver of a communication are excluded." It is precisely this "supplemental" distribution of axiological, cultural values that creates the possibility of an actantial typology of genres. In narrative, then, cognition and culture coincide, and the structuration of literary narrative into narrative genre will make this coincidence evident.

The Structure of Actants

If the relation between sender and receiver articulates the situation of linguistic activity, then the other actants establish the message as a simulation of the world by superimposing upon the serial messages of language structural relationships beyond the confines of the sentence that can be apprehended as a "meaningful whole" (Greimas 1983b: 59). In literary studies such "wholeness" is what is defined (more or less) by genre. Before turning to literary genres, however, we should look once more at the logic of Greimas's actants in relation to Lévi-Strauss's matrix. "Our own analytical experience," Greimas writes,

> as well as that of other semioticians' has shown that, to account for even slightly complex texts, it is necessary to envisage the possibility of the splitting up [*éclatement:* "exploding"] of any actant into at least four actantial positions that we can present by using terminology proposed by Jean-Claude Picard:

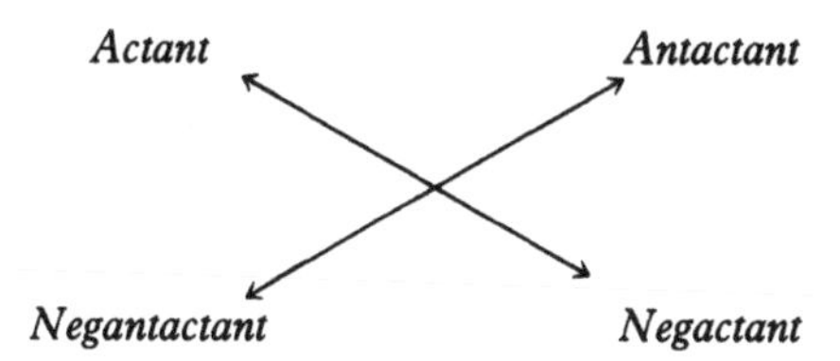

(1988: 44–45)

Beginning with the subject, we can represent this scheme with Greimas's actants from which the sender and the receiver are excluded (see Figure 5).

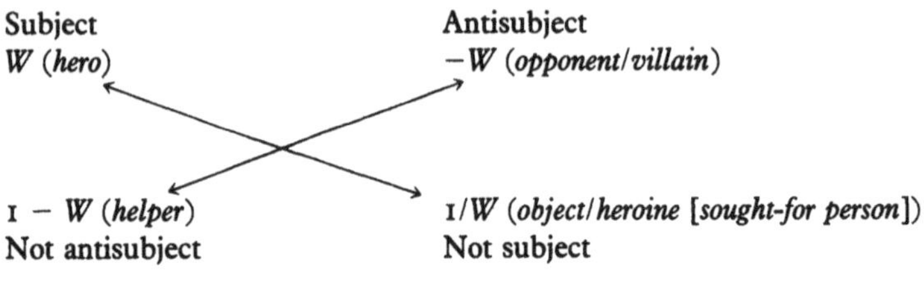

Figure 5

If we adopt the actants named in parentheses as four basic character roles, we note that Lévi-Strauss's matrix, inscribing these actants defined relationally (W, $-W$, $1/W$, $1 - W$), suggests four modes of discourse and four actantial structures of narrative genres.

We are designating these four modes heroic, agonistic, synecdochical, and ironic, and the four types of narrative melodrama, tragedy, comedy, and irony. These generic terms echo the categories Northrop Frye designates in the "Theory of Myths" in the *Anatomy of Criticism,* and the nature of actantial modes we are describing resembles Frye's "Theory of Modes" (1957: 158–60, 33–34). In the *Anatomy,* for example, Frye describes modes in terms of the quality of the actants—what Propp calls the "spheres of action" of dramatis personae. In the *mode* of romance the hero is superior in "degree" to "man" and "nature," while Frye describes the *genre* of romance in terms of its formal relationship to theme. The distinction between mode and genre, implicit in Frye's work, is explicit in Alastair Fowler's *Kinds of Literature*. "The terms for kinds," he writes of genres, "perhaps in keeping with their obvious external embodiment, can always be put in noun form ('epigram'; 'epic'), whereas modal terms tend to be adjectival" (1982: 106). To determine a mode by the power of action of the hero describes the modal *quality* of a discourse; to determine a genre by an "external embodiment" provides *formal* distinctions.

Later we will see how Greimas's actantial terms allow the structural classification of both the modality and the formal, "external embodiment" of narrative forms, but here we will explore the important difference between Frye's categories and actantial categories. This difference is most apparent in the fact that Frye's categories are based on content rather than the positions of logical relationships. Frye determines each genre by particular mythic content and each mode by positive attributes of the personae measured against particular qualities in the world. In this Frye is *formalist* rather than *structuralist*. He is situating himself within the categorical distinction between epistemology and methodology as an "epistemologist" without questioning, as Continental structuralism and Anglo-American cognitive science do, that very distinction. In this, he seeks to organize content formally rather than to apprehend a logical structure of its embodiment. As Lévi-Strauss says in his critique of Propp, "*Form* is defined by opposition to content, an entity in its own right, but *structure* has no distinct content: it is content itself, and the logical organization in which it is arrested is conceived as a property of the real" (1984: 167). Tzvetan Todorov notes that, in Frye,

> the *structures* formed by literary phenomena *manifest themselves at the level of these phenomena*—i.e. these structures are directly observable. Lévi-Strauss writes, on the contrary: "the fundamental principle is that the notion of social structure is not related to empirical reality but to the model constructed according to that reality." To simplify, we might say that in Frye's view, the forest and the sea form an elementary structure; for a structuralist, on the contrary, these two phenomena manifest an abstract structure which is a mental construction and which sets in opposition, let us say, the static and the dynamic. (1975: 17)

By "abstract structure," Todorov means *cognitive structures*, the kinds of *logical* relationships that Greimas and Lévi-Strauss abstract and articulate in terms of the positions of the semiotic square and the Boolean matrix.

In these terms, actantial categories are cognitive structures. They are positions determined by the relationship between actants. Moreover, the logical organization of actants—the "abstract" logical relationships between actantial roles—"arrests" an organization of "content," a typology of modes and genres. It does so because actants (and the genres and modes they organize) are located between the abstract and the concrete, just as structure for Lévi-Strauss—and cognition itself, generally conceived—mediate between content and logic. The heroic mode, for example, is defined by the *position* of the hero in relation to other actants, and the narrative genre of melodrama "discursivizes" (as Greimas says) that position. The hero faces an opponent or series of them, defeats them, and achieves revenge and glory. In the agonistic mode the hero is positioned differently so that, discursivized in tragedy, the hero's struggle with his opponent proves fatal to both because he shares some quality or "flaw" with the opponent. The other modes define the hero-subject through repositionings that imply different generic forms—different acts of cognitive apprehension. In comedy hero and heroine are equal partners in the struggle with the opponent. They win and marry. In irony the hero lives with the realization of his failure, which is also the realization of self-division, a hero who is not a hero.

The structural logic of these actantial positions becomes clear with the superimposition of Lévi-Strauss's vertical column on Greimas's square inscribed in Figure 5. Hence, in Figure 5, *W* is the subject-hero. $-W$ is the opposite of the hero: the antihero or villain. The relationship between *W* and $-W$ is a *contrary* relationship of reciprocal presupposition. $1/W$ is the inverse of the hero: the heroine or sought-for person. The heroine is *categorically* different from the hero, and not simply a female hero. The relationship between *W* and $1/W$ is a *contradictory* relationship. $1 - W$ is the complement of the hero, the helper. The relationship is a *complementary* relationship of simple presupposition.

Hero ≈ *Villain* (W ≈ −W). Both villain and heroine are in some sense in opposition to the hero, but the principle of opposition is different. Villain and hero are in contrary opposition in which the opposition is based on the absence or presence of some quality. "Hot" and "cold" are such an opposition (cold is the lack of heat, heat is only conceivable in relation to cold), and neither element is conceivable without the other. The hero and villain are opposed on the basis of goodness as opposed to evil, which is the lack of goodness. Still, as we will see, the relative goodness and evil of the hero and opponent vary widely according to genre as the position of these actants in relation to the other actants changes.

Hero ≈ *Heroine* (W ≈ 1/*W*). The hero and heroine, on the other hand, are in contradictory (or equipollent) opposition. Traditionally, the hero and the "sought-for person" are opposed on the basis of sexuality, male vs. female, terms which are equivalent but categorically different. Greimas marks this categorical difference in an analysis of Maupassant's "Piece of String" in which he notes that "the category of sex" in the story divides "two distinct types of doing." Men shop and bargain while women sell so that "masculine doing is for the most part verbal, whereas feminine doing is an almost entirely somatic one of an economic nature." "We can see that such a distribution of activity according to the classes of sex," Greimas continues, "is not pertinent to the 'referential' plane and that to account for this we must try to find another pertinence within the semantic organization of discourse" (1989a: 622).

That is, sexual difference is neither an absolute and positive "content," nor simply a "formal" arrangement imposed on experience to make it comprehensible. Rather, as Lévi-Strauss says, it is the "logical organization" in which "content" is "arrested and conceived as a property of the real." As a "logical organization," it can be inscribed within the elementary structure of signification. This is why Greimas defines the "heroine" simply as an "object" of desire, not even a "sought-for person," but a sought-for "good" (1983b: 232). But as "a property of the real," sexual difference has palpable social and semantic effects, effects which cannot simply be "formally" eliminated by changing the terminology. Throughout our analysis we use the traditional term "heroine" as well as more abstract designations because that traditional employment will often clarify the "real" meaning-effect of this actantial role in narrative genres. But even here the term designates a *structural position* rather than a sexual stereotype, even when that position is inhabited by a woman, as it often is in Western narrative.

Hero ≈ *Helper* ($W \approx 1 - W$). Finally, the *contrary* relationship between the hero and villain and the *contradictory* relationship between hero and heroine are different from the relationship of hero and helper. Unlike both of these, the helper simply presupposes and *complements* the hero who helped, and it is possible to conceive of heroes without helpers. Greimas describes the helper in terms of abstract "forces in the world" more or less realized, so that in some discourse the helper can be seen simply as the "power of acting" of the hero himself (1983b: 206, 280).

Actantial Modes

When Lévi-Strauss's matrix is read vertically, column by column, the relationships among the actants change, and with those relational changes the literary modes we have designated define themselves. That is, literary modes define themselves when the matrix is followed from top to bottom (W, $-W$, $1/W$, $1 - W$). The first begins with W (the actantial "hero" in our investment) and defines, or structurally articulates, the *heroic* mode. The second begins with $-X$ (the "villain") and defines the *agonistic* mode. The third begins with $1/Y$ (the "heroine") and defines the *synecdochical* mode. And the fourth begins with $1 - Z$ (the "helper") and defines the *ironic* mode. In this way each symbol in the matrix (W, X, Y, Z) defines an actantial role, and each column describes a particular mode of narrative defined in terms of the relationships defining those roles (see Figure 6).

Modes			
heroic	agonistic	synecdochical	ironic
↓	↓	↓	↓
W	$-X$	$1/Y$	$1 - Z$
$-W$	$1/X$	$1 - Y$	Z
$1/W$	$1 - X$	Y	$-Z$
$1 - W$	X	$-Y$	$1/Z$

Figure 6

Actantial modes, then, simply classify texts according to their focus on one or another of these basic actants. By "focus," we mean that in these narratives different actantial roles assume the position of Greimas's (syntactic) subject. Roles are determined by binary relationships, and focus is determined by a different set of relations. When we described the structure of actants in Figure 5, we had, in fact, defined a particular mode of narrative, the *heroic* mode. In narratives that "focus" on the hero (W), it is the hero standing by himself, as it were, that defines the relationships between him and the other actants and, thus, the quality of those actants and the value of the narrative. Such narratives are extremely common. Most tragedies center on a tragic hero, although a few (e.g., *Richard III*) focus on a villain. Epics such as *The Odyssey,* romances like *Parsifal,* and modern thrillers such as *Raiders of the Lost Ark* are all hero-centered stories, as are most narratives of initiation. What distinguishes these stories is the reciprocal presupposition of hero and villain. Propp's wondertales fit the scheme of Figure 5, and the fact that he (and later Lévi-Strauss) used such stories to attempt morphology and structuration of narrative seems appropriate from the fact that its relational structure is privileged: both vertical and horizontal readings of Lévi-Strauss's matrix position the actants in a way that corresponds to the semio-logical categories of Greimas's semiotic square. It is for the same reason that we used this narrative mode to articulate the structure of actants.

In the *agonistic* mode of narrative, the opponent (X) is the focal point. Here, in the relationship between the opponent and the hero, the hero presupposes the opponent (see Figure 7).

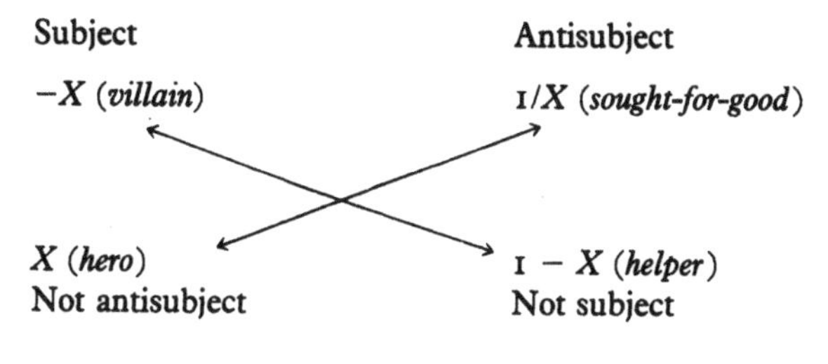

Figure 7

Such stories would include tragedies in which the hero is or becomes the villain (e.g., *Macbeth, Richard III*). Comedies such as *The Merchant of Venice, Volpone,* and *Miles Gloriosus* in which the blocking characters (opponents), Shylock, Volpone, and Pyrogopolynices, are more important than the lovers are modally agonistic. Unlike heroic narratives, agonistic narratives depict a hero who shares some negative quality with the villain or opponent.

The *synecdochical* mode of narrative focuses on the heroine (Y) (see Figure 8).

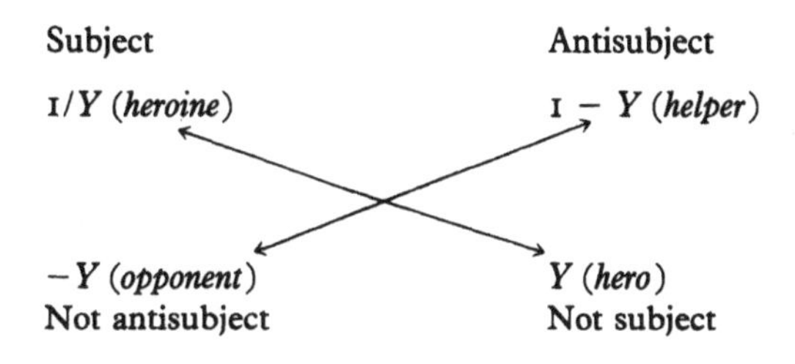

Figure 8

These narratives focus on the heroine as a sought-for good, and not simply as a female hero. By this criterion, *The Duchess of Malfi* and *Mrs. Dalloway* would not qualify, but narratives in which damsels-in-distress are rescued, such as the *Perils of Pauline,* do. *Clarissa* is a novel that focuses on the heroine as heroine rather than female hero. *Pamela* is a heroine narrative, and *Shamela* is a parody precisely because it transforms the heroine into a female hero. Gothic romances told from the point of view of the terrified heroine are also examples. What distinguishes these narratives is that the heroine is presupposed by the opponent, who is quite often a family member (e.g., New Comedy) or a lover (e.g., *Clarissa*). A play such as *A Doll's House* might be seen as either a heroic or a synecdochical narrative, but we prefer to classify it as modally synecdochical because of the subordinate role Nora plays throughout the drama—even the title makes her into a "good," namely a doll—even though she becomes heroic in the end. *A Doll's House* is an example that helps us articulate the structural distinctions we are making. Nora is a "hero" whose sought-for good is not a "heroine" but the quality of heroism her society denies her but grants to her opponent, her husband. But he is the opponent only insofar as he is married to Nora, that is, only by presupposing the heroine.

There are narratives in which the sought-for good (Greimas's "object") is not the heroine but something else. In these narratives the sought-for good is seldom in an equipollent relationship with the hero, and they are usually modally heroic. Examples of works in which the sought-for good does share dominance with the hero to some extent are those in which that good takes on *symbolic* significance. These include Grail stories such as *Parsifal,* modern quest stories like *The Treasure of the Sierra Madre* and *Raiders of the Lost Ark*. In calling this mode *synecdochical,* we are following Kenneth Burke's description of the four major tropes in which metonymy is "reductive"—the either/or of the agonistic mode—while synecdoche is "representative"—the inclusive both/and of symbolic and comic modes of discourse. For Burke metaphor is the trope of "perspective" which, like the heroic mode, measures all relationships from the vantage of the hero-subject (1969: 503–17; see also Kellner 1981: 14–28).

The final actantial mode is *ironic,* comprising narratives in which the helper (*Z*) is the focal character (see Figure 9).

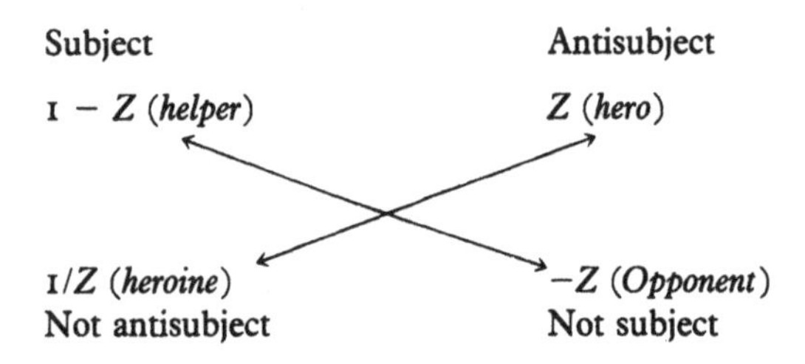

Figure 9

"Rumpelstiltskin" is a good example of a folktale in the ironic mode because it demonstrates the way the helper is presupposed by the heroine. The little man is more important than the heroine. By "important," we mean that his narrative action *functions* more explicitly and more centrally in the unfolding narrative than that of the young woman whom he helps weave straw into gold. Greek and Roman New Comedies which feature clever slaves can be classified as helper narratives in the same way. Davos in *The Woman of Andros* is an example of a helper who is more important than the hero (who is unimpressive) and the heroine (who never appears on stage). The clearest examples of the ironic mode are stories in which the helper appears as the "sidekick." Famous sidekicks include Sancho Panza, Huck's Jim, Falstaff, Nick Carraway, and even the Gabby Hayes/Andy Devine figures in Hollywood films. In most cases this sort of helper is subordinate to the hero (that is, these are heroic narratives [Figure 5]), but Sancho Panza is in a relationship of reciprocal presupposition with Don Quixote, and Falstaff overshadows Hal in the way the hero overshadows the villain in the heroic mode. In fact, Falstaff in many ways identifies the helper actant and helps define the identification of the actantial role of helper with the ironic. Burke writes that he "would consider Falstaff a gloriously ironic conception because we are so at one with him in his vices, while he himself embodies his vices in a mode of identification or brotherhood that is all but religious. Falstaff would not simply rob a man, from without. He *identifies himself* with the victim of a theft" (1969: 515). To put this in "positional" or structuralist terms: Falstaff defines the helper as an actant in an equipollent relationship with the opponent.

Actantial Genres

If actantial modes are defined by the relationships existing between actants—if they are a function of the "focal" actant defined relationally—then actantial genres, like the genres described by Frye and Fowler, are defined *formally,* in relation to the criteria for what Fowler calls their "external embodiment." Now the actants that "embody" such external factors in Greimas's structural semantic narratology are, as we have seen, the sender and the receiver. The formal relationship that defines actantial genres, then, is the syncretic relationship between one of the "internal" actants and the receiver. It is the receiver that is left with cultural values at the end of a narrative program, and those values seem to transcend the very narratives that embody them. In the examples Greimas gives in *Structural Semantics*, the subject-hero of the Russian wondertale is the receiver of the sought-for good—the princess—while in the narrative of a capitalist, it is the object—the economic enterprise—that is syncretized with the receiver (1983b: 204, 210). In both cases the sought-for good embodies transcendental cultural values: receiving the princess, the hero "receives" the state. The enterprise itself receives the values of the sender which Greimas designates as the "economic system."

What is most impressive about Greimas's actantial analysis is that it allows the structuration of "external," cultural criteria. It creates a relationship between cognition and culture. The actantial category sender vs. receiver inscribes within the "systematic structure" of discourse a category that "explodes" that systematicity. Unlike Hjelmslev's systematic linguistics (1961: 79), Greimas's semantics attempts to inscribe "content-substance" within language within the cognitive project of his semiotic analysis of genres, the classification of the thematic-formal aspects of literary discourse. That is, genre categories attempt what semiotics and cognitive science attempt, to articulate structures of signification, of apprehended meaning. In this double reading of narrative theme and logical form—Lévi-Strauss calls it "the double aspect of time representation" (1984: 183)—Greimas is following Lévi-Strauss, who inscribes his "structural semantics" in the Boolean matrix. That is, the matrix can be read horizontally as well as vertically, and here, rather than relationally defining actants, it defines narrative genres. Each horizontal line represents a different genre: the top melodrama, then tragedy, then comedy, and then irony (see Figure 10).

Genres		Modes			
		heroic	agonistic	synecdochical	ironic
		↓	↓	↓	↓
melodrama	→	W	$-X$	$1/Y$	$1-Z$
tragedy	→	$-W$	$1/X$	$1-Y$	Z
comedy	→	$1/W$	$1-X$	Y	$-Z$
irony	→	$1-W$	X	$-Y$	$1/Z$

Figure 10

The top line, melodrama, replaces Frye's category of romance. Our objection to "romance" is not simply that it describes the mode of a work rather than a formal genre. More important, Frye believes "romance" implies particular content: he groups works as diverse as *Beowulf, The Faerie Queen, Pilgrim's Progress,* and *Uncle Tom's Cabin* because they exemplify some form or another of the quest myth. Genre, on the other hand, is an abstract semantic class of works—a cognitive or semionarrative classification—whose semantic structure must define a class independent of particular content. Rather than being defined by a positive content such as the presence of a "quest," it is defined by the logicosemantic relationships that inform its elements.

Melodrama, we argue, best describes that *form* of literature that embodies the heroic mode. The OED defines "melodrama" as a "dramatic piece characterized by sensational incident and violent appeals to the emotions, but with a happy ending." Peter Brooks adds that it exemplifies an "aesthetics of excess" characterized by "heightened and polarized words or gestures" (1976: 202). A popular definition in the *Dictionary of Literary, Dramatic, and Cinematic Terms* calls it a "drama wherein characters clearly virtuous or vicious are pitted against each other in sensational situations filled with suspense" (Barnet 1971). These definitions are actantial in that they define the melodramatic as we defined the heroic: as a confrontation of reciprocal roles in which, from the perspective of the hero, the virtuous emerge victorious, and the vicious are crushed. Such a modal definition can be formalized in actantial terms, in terms of the syncretism of one of its actants and the receiver. In melodrama the subject-hero is syncretized with the receiver. This definition would extend the use of the term "melodrama" to include fiction and narrative verse as well as drama, and include one of the oldest extant examples, *The Odyssey,* as a prototype of the form: the wily Odysseus (W) triumphs over or escapes from a series of opponents—Polyphemus, Circe, the Sirens, Scylla and Charybdis—before the climactic confrontation with the collective actant which serves as the major villain, the suitors ($-X$). With the aid of Eumaios, the aged swineherd who fills the role of helper ($1 - Z$), Odysseus crushes the suitors and "receives" the heroine, Penelope ($1/Y$). In these terms, the Russian wondertale as Propp describes it—and the mythic discourse Lévi-Strauss describes throughout his career—are melodramas. In fact, as Alan Dundes notes in the preface to *Morphology of the Folktale,* Propp's morphology accurately describes the second half of *The Odyssey* (Propp 1968: ix). In the line that represents melodrama—W, $-X$, $1/Y$, $1 - Z$—the hero of melodrama is positioned as W. This position is unmarked because the hero is properly heroic and triumphant. His triumph is embodied in the fact that he is also the receiver. In more general terms, when the matrix is read horizontally—that is, when it is read *generically*—the actant that is unmarked is the actant that is syncretized with the receiver.

Melodrama has the sharpest differentiation between characters of any of the four actantial genres. The heroes are distinctly heroic, and the opponents, defined against the hero, are "self-evidently" villainous and despicable. It is precisely the qualities of clarity and simplicity that allow Lévi-Strauss and Greimas to reconceive Propp's description of a subset of melodrama, the Russian wondertale, in structural terms. In fact, although Odysseus is an exception, the characters of melodrama are usually stereotypes, and this is an important reason that few melodramas are considered serious literature. The heroes are the essence of virtue, although what constitutes virtue differs dramatically in different periods. Purity in the hero in one era is constituted by forms of innocence, whereas in another era it is constituted by worldly wisdom. What is most important, however, is that the hero embodies and receives in the end the cultural value, the sought-for good.

All the examples of melodrama focus on the hero who defines the other actants in terms of himself. His degree of heroism defines the degree of villainy of the opponent; his worthiness defines the value of the heroine or the sought-for good; his power defines the supplementarity of the helper. The opponent in melodrama, then, is the contrary to the hero, the essence of evil defined as the polar opposite of heroic virtue. He is a nightmare character who crystallizes our worst fears. Sometimes he is merely a human deficient in humanity, like Simon Legree or Lovelace. Sometimes he is a monster from the depths of hell, like Dracula or the shark in *Jaws*. The heroine of Western melodrama is beautiful and pure, beset by evil, often explicitly or implicitly sexual. Whether the monster is a human who wants to shame her, like Iachimo in *Cymbeline,* or a monster who wants to suck her blood in *Dracula,* the effect is the same. Often the dark figure grabs her and takes off over the rooftops, as in *The Cabinet of Doctor Caligari* or *King Kong,* and this iconic scene (helpless heroine, dark villain, hero watching in frustration below) is the essence of Western melodrama. In the end, the heroine, like the hero, embodies the cultural values of the work, and her receiver-hero obtains and realizes them. The helper of melodrama, when the character exists beyond the realized power of the hero himself, is a comic foil to the hero. In superficial terms at least, Falstaff appears to have set the pattern for modern helpers. He is funny, fat, bibulous, and cowardly.

Most melodramas are popular works of the mass media or, as in Propp, of folklore. Melodrama, as the unmarked narrative form, finds the widest articulation in popular culture. Still, melodrama also inhabits canonical literature. Dickens favored the form, and *Oliver Twist* and *David Copperfield* are novels in which young heroes overcome formidable obstacles eventually to triumph. Shakespeare's *Henry V* is a classic example of melodrama, pitting the noble Henry and the brave English against the boastful Dauphin and the despicable French. The showdown at Agincourt is the classic climax of melodrama. Both parts of *Henry IV* are melodramas, although they vary the pattern somewhat. In *1 Henry IV* the moral distinction between Hal and the people loyal to Henry IV, on the one hand, and Hotspur and the rebels, on the other, does not seem to be a sharp one to us today. Hotspur certainly is not evil like the villains of later melodramas. Nonetheless, when Worcester and Vernon keep Hotspur in the dark about the king's offer of a truce and pardon, it becomes clear that the rebels' cause is wrong. Furthermore, Hotspur's speech on honor, which sounds so elevated today, is less impressive in its context. Worchester, who hears it, remarks: "He apprehends a world of figures here, / But not the form of what he should attend" (I.iii.209–10). In any case, *1 Henry IV* climaxes with the traditional confrontation of melodrama. Hal tells Hotspur that England is not big enough for the both of them—"Two stars keep not their motion in one sphere; / Nor can one England brook a double reign" (V.iv.65–66)—and then kills him. *2 Henry IV* is a weaker example of the form—it is a modally "agonistic" melodrama—but Shakespeare still follows the pattern of two opposed groups in a relationship of reciprocal presupposition.

The next horizontal row of the matrix structurally articulates the genre tragedy, which formally embodies the agonistic mode. Traditionally the term "tragedy" has been used primarily for drama, but as a narrative genre we may use it to include novels such as *Lord Jim* and *Tess of the D'Urbervilles*, romances like *Le Morte D'Arthur* and *Le Chanson de Roland*, and examples of minor narrative genres like the epyllion *Venus and Adonis*, and Frost's short narrative poem "The Death of the Hired Man." It is a critical commonplace that characters in tragedy are usually more complex than those in melodrama. Examining the matrix may suggest a reason for this phenomenon (see Figure 10). As we move down the rows from melodrama to tragedy to comedy to irony, the differences between the actants become less distinct. The opposition between hero and opponent, based on virtue, is strongest in melodrama, where the hero is distinctly virtuous and the villain obviously vicious. In tragedy the hero is a tragic hero, a figure Aristotle describes as a "man who is not eminently good and just, yet whose misfortune is brought about not by vice or depravity, but by some error or frailty" (1989: 69). The phrase "error or frailty" represents the critical disagreement as to what Aristotle meant by *hamartia*, but whether he meant a constitutional blemish or the commission of a grievous sin (what *hamartia* means in the New Testament), the tragic hero is more complex than the hero of melodrama because he is a combination of strengths and weaknesses, virtues and vices. He is a *marked* figure ($-W$). Here we can see that the two parts of *Henry IV* are melodrama disguised as tragedy insofar as Hal is *apparently* weak. His transformation, like that of the hero of the Russian wondertale Propp examines, is what Greimas calls the "revelation of the hero." While there are tragic villains who approach the evil of those of melodrama—Iago comes to mind—more often the opponent is complex and ambiguous, so that in Antigone and Creon, Romeo and Tybalt, Antony and Caesar, and even Hamlet and Claudius, we have a protagonist and antagonist rather than a hero and villain. Even the exception, Iago, helps define Othello as the most "heroic" of tragic heroes and the tragedy *Othello* as modally heroic.

The line for tragedy on the matrix reads $-W$, $1/X$, $1 - Y$, Z. If we observe the position of the terms in the lines as we descend through the rows, it is apparent that they rotate, like players on a volleyball court. The hero, which was represented by the unmarked term in the top line (melodrama), is now the contrary. The villain is no longer the contrary to the subject-hero but is in contradictory opposition to the unmarked term, now inhabited by the helper. The relationship of the first three terms indicates that the hero has declined from his flawlessly heroic state in melodrama. No longer are he and the opponent in a relationship of contrary opposition—one good, one not good—but in a relationship of functional equipollence, opposed in what they want or stand for, like Antony and Caesar, Coriolanus and Aufidius, Jason and Medea, Antigone and Creon. The heroine in tragedy often assumes more importance than she has in Western melodrama, which may be indicated by her designation as complement, $1 - Y$. Tragedies are often stories of tragic pairs: Tristan and Isolde, Hero and Leander, Pyramus and Thisbe, Romeo and Juliet, Antony and Cleopatra. We cannot think of a melodrama in which the hero shares the billing with the heroine.

The helper is represented by an unmarked term, and this indicates the syncretic fusion of helper and receiver. At the end of tragedy, unlike melodrama, it is the helper and not the hero who receives the transcendental cultural values. Horatio refrains from killing himself to help reestablish the state in *Hamlet;* Kent does the same in *King Lear,* and Creon does so in *Oedipus*. Very often, moreover, it is the helper, like Horatio, who lives to tell the tale: Ishmael in *Moby-Dick,* Marlow in *Lord Jim* and *Heart of Darkness,* Nick in *The Great Gatsby.* What the helper receives is the ability to speak the language that can recognize and convey cultural values even when everything is seemingly destroyed. (If the receiver in literary genres can recognize and convey cultural value, a culturally situated sender, as we suggest in the following chapter, can encompass and convey cultural value.) Tragedy needs a helper precisely because it needs both to destroy what is—and the values "sent" to what is—and to assert its values *discursively.* With the narrative genre of tragedy, as opposed to melodrama, the *cultural* (rather than *individual*) nature of cognition emerges. In their *formal* lack of cultural cognition is another reason so few melodramas are considered "serious" literature.

The third horizontal line on the matrix, $1/W$, $1 - X$, Y, $-Z$, structurally articulating the genre comedy and formally embodying the synecdochical mode, furthers this emergence of cultural cognition. Like "tragedy," "comedy" is a term that traditionally has referred primarily to drama but may be extended to other Western narrative forms as well. The pattern established by Menander's New Comedy has been the dominant comic formula ever since the fourth century B.C.: a boy wants a girl but cannot have her because of some obstacle, usually parental disapproval. By intrigue or luck, the lovers overcome the obstacle and eventually live together. This formula was adopted by Plautus and Terence and then disseminated throughout Europe. Very few comic dramas in any subsequent period have used any other pattern, not simply because of this dissemination, but because of the semantic logic of the form. In Shakespeare's adaptation of the New Comedy formula, sometimes the lovers are the focus of the play, as in *The Taming of the Shrew* and *Much Ado about Nothing*—both "synecdochical comedies"—and sometimes the opponent is, as is Shylock in the "agonistic comedy" *The Merchant of Venice.* At times the lovers are very minor characters who make an obligatory presence for form's sake but play little part in the drama, as in the "ironic comedy" *The Merry Wives of Windsor* in which the story of Anne Page and Fenton provides the skeletal structure, but Falstaff provides the flesh and blood, so to speak. Ben Jonson's comedies of humors and Restoration comedies of manners follow the New Comedy pattern, and, for all his insistence on the primacy of content over form, Shaw uses the very same pattern in *Major Barbara* and *Man and Superman,* among other plays. Broadway and Hollywood made this pattern the basis of musical comedy, and in our day it remains popular in the works of playwrights such as Neil Simon.

Tom Jones, which Fielding called a "comic epic in prose," has many melodramatic moments but is essentially an actantial comedy—a "heroic comedy"—constructed on the formula of New Comedy. After surmounting a series of obstacles, Tom marries Sophie. What distinguishes this from a melodrama is precisely the nature of the marriage at the end. If Sophie is the sought-for good, Tom can hardly be said to inherit the state when he marries her, and if she "embodies" wisdom, it is domestic rather than transcendental wisdom. Such domesticity helps define comedy. On the line of the matrix articulating comedy, the heroine becomes the unmarked term. The designation of the heroine as the receiver is appropriate for a form that so often culminates in marriage. More important, the softening of the oppositions that occurs in the descent from melodrama to tragedy continues in the move from tragedy to comedy. In tragedy there are scattered examples of hero and heroine sharing the focus of the drama—*Romeo and Juliet, Antony and Cleopatra*—but in comedy this is the rule rather than the exception. Furthermore, in most comedies the opponent is no longer the villain; instead, it is commonly the parents. One exception is Volpone, a villain who acts as the opponent, but it must be noted that *Volpone* is modally heroic and very close to melodrama.

The hallmark of comedy, in opposition to melodrama (comedy is, in fact, the *contradiction* of melodrama), is the roughly equal status of the characters. In both melodrama and tragedy, the hero towers over the other characters and dominates the action and the audience's attention. In comedy the hero and heroine are a pair; the opponent plays a crucial role of opposition, and the helper is often of equal or even greater importance than the lovers, as in the case of Roman New Comedy, where the clever slave who contrives the intrigue is the center of interest in the play. Nevertheless, the heroine is the receiver. If tragedy asserts the *cultural* (or public-cultural) value of humanity in the destruction of the hero—a value that is verbalized (or "discursivized") by the helper-receiver—then comedy perpetuates the *natural* (or domestic-cultural) value of humanity in the renewal of generation embodied in the heroine-receiver. She receives the hero not as statesman and warrior but as husband. *King Lear* is the tragedy of growing old and having to relinquish the state, the supreme contrary to melodrama. *The Playboy of the Western World,* whose "plot" in broad outlines repeats that of *Lear,* is the comedy of Christy's positioning himself to find a wife who will receive him. In this contrast we can see that generic distinction, as we have argued, is a function of the receiver-actant. In these terms *Tom Jones* is a comedy precisely because Sophie receives Tom, not in order to articulate and pass on cultural values in discourse as the helper-receiver does in tragedy, nor to articulate and embody human power as the hero-receiver does in melodrama, but to disarticulate the sharpness of actantial distinctions in a family constellation. It is a comedy because marriage softens the distinction between self and other in domestic culture.

The last line of the matrix, $1 - W, X, -Y, 1/Z$, formally articulates the ironic mode. In the genre of irony the hero faces obstacles, fails to overcome them, and usually does not die but is trapped and must live with his defeat. In irony, however, unlike the other narrative genres, no helper of the hero is fused with the receiver. The values the hero embodies or seeks are "received" by neither the heroine, nor hero, nor helper but by the opponent, the antihero.

Irony is perhaps the most common form in modern fiction. Virtually all of Joyce's stories assume this narrative form, for instance. In "The Boarding House," a possible hero, Mr. Doran, believes he is seducing the heroine, Polly Mooney, while she is silently trapping him into a marriage he does not want. The story ends with his realization of the fact that he is condemned to a lifetime with a woman he may or may not love, and he has a vague sense of "being had." (In *Ulysses* we see Doran again, roaring drunk and hiding from his family.) Here is irony disguised as comedy: the receiver is not the heroine, Polly, but rather the opponent, Mrs. Mooney, Polly's mother and the butcher's daughter. In "A Painful Case" James Duffy comes to the realization that "no one wanted him; he was outcast from life's feast" (1967: 117). The last line of the story, "he felt that he was alone," could serve as the last line for "Counterparts," "The Dead," "Eveline," and indeed not only all the stories in *Dubliners* but any example of the genre of irony. In irony the focus shifts from hero to helper, who is usually an ineffectual parody of a hero. As in Joyce's epiphanies, there is really no story to tell—no narrative program—but rather sundry "epiphanies" that can or cannot be taken as revelations. Irony is, in effect, an anticognitive narrative in which neither the cognitive significance nor the narrative itself is clear.

If, as we have suggested, narration and cognition are binary opposites, then the narrative genres we have elaborated, following Propp, Lévi-Strauss, and Greimas, are "contradictory" to narrative insofar as genre resituates particular narratives on the level of system. Irony, then, is in a "contrary" relationship to narrative genres, neither narrative nor understanding but simply dissociated events (see Figure 11).

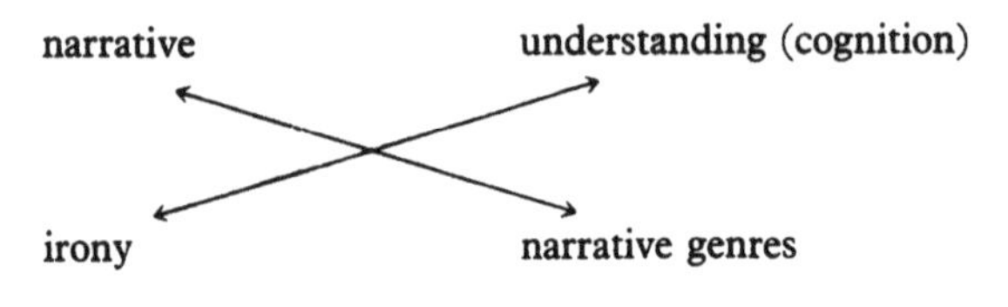

Figure 11

Figure 11 also suggests a typology of genres. Melodrama emphasizes or foregrounds narrative—hence Propp's and Lévi-Strauss's focus on this narrative genre in structuring narrative as such. Tragedy focuses on understanding—Aristotle's *recognition*. Comedy focuses on the wholeness and harmony of comprehension—the "meaningful whole" that is the explicit focus of Greimas's account of narrative (1983b: 59) and the "harmony" that is the implicit criterion of Whitehead's definition of science. Irony focuses on fragmentation, perhaps most terribly articulated in Paul de Man's ironic reading of Shelley where he asserts: "*The Triumph of Life* warns us that nothing, whether deed, word, thought, or text, ever happens in relation, positive or negative, to anything that precedes, follows, or exists elsewhere, but only as a random event whose power, like the power of death, is due to the randomness of its occurrence" (1984: 122). Such a reading, like the hypostatization of signs that Latour critiques in semiotics (1986: 26)—though in a very different register—effectively ignores the cultural meanings, power, and goods that Greimas takes such pains to include within his cognitive semantics. (Such irony, though, can turn back on itself and, as we suspect it does in de Man, situate "irony" itself as a cultural value.)

Within the typology of genres, then, irony is *contrary* to comedy as it is *contradictory* to tragedy. Rather than sharp distinctions between actants, almost all distinctions are erased so that it is difficult to distinguish between heroine and opponent, opponent and helper, helper and hero. In "A Painful Case," the opponent, Mr. Sinico, is also a kind of ineffectual helper who "receives" the sought-for good in the end. "Captain Sinico," we are told, "had dismissed his wife so sincerely from his gallery of pleasures that he did not suspect that anyone else would take an interest in her" (1967: 110).

The striking fact of irony is that it so often approaches, as does parody, other genres. In this we can see the "narrated" breakdown of cognition. The narratives of *Dubliners* approach not only the comedy of courtship, but even the melodrama of the quest romance. At the end of "The Dead," for instance, Gabriel might become the receiver of Greta (as, in his fantasies, he once had been) and achieve "heroic" stature in a westward quest. Or his opponent, Michael Furey, may be the receiver (as *he* once had been), and Gabriel remains the "pitiable fatuous fellow" he sees in the mirror. The fact is that irony gives no criteria by which to choose between these alternatives. It fails to do so not simply because, as Colin MacCabe has argued, Joyce eschews the metalanguage that authoritatively interprets the action in traditional fiction (1975: chap. 2). Rather, irony gives no actantial criteria precisely because it replaces the *action* of narrative with nonaction. For example, "The Dead," like *Ulysses* (and like *The Sun also Rises,* which ends with the famous exchange "'Oh, Jake', Brett said, 'we could have had such a damned good time together.' . . . 'Yes,' I said. 'Isn't it pretty to think so.'"), can only be authoritatively interpreted *hypothetically,* depending on what happens next, that is, after the end of the narrative. Irony, then, is a narrative genre that replaces narrative with a kind of contrary-to-fact subjunctive. It "explodes" narrative form so that the "functional" actantial analysis of narrative and genre approaches the qualitative analysis of modes. Here it is clear why we have retained the "modal" name for this genre, "irony."

Perhaps an example of irony's approach to tragedy—irony disguised as tragedy—will clarify this point. Boswell's *Life of Johnson* can be read as tragedy: the helper Boswell receives and recites the narrative, just as Nick does for Gatsby, and Ishmael for Ahab. For years the *Life* has been read as Johnson's tragic encounter with life in a corrupt world. Yet the receiver can also be understood as the opponent, Death, whom Johnson so feared throughout his life, and the whole can be read as the (ironic) story of Johnson being overwhelmed by life, of living-for-Death. Here again, the choice can only be decided by what happens next: by the quality of Johnson's afterlife. In a conventionally faithful age—or, in the case of Matthew Arnold, a desperately faithful one (see, for instance, Miller 1985)—Johnson is the figure of modern tragedy. In an agnostic age, he is the figure of irony.

Another irony approaching tragedy is *Waiting for Godot.* In that play Vladimir and Estragon are the joint hero. There is no heroine, and the opponent, if there is one, could only be life or fate (but "fate" conceived as nonsensical, nondiscursive, not susceptible to cognitive apprehension). Godot is the absent helper (confusingly close to the absent sought-for good), and it is precisely the absence of the helper—who is the receiver in tragedy (or the absent sought-for good, who is the receiver in comedy)—that transforms tragedy into irony, discourse into repetition, cognition into bewilderment. Without a helper for the hero, cultural values—values beyond the individual—are impossible (which is why irony is contradictory to tragedy). *Godot,* like Joyce's *Dubliners* and, perhaps, Boswell's *Life,* can be described as a nonagonistic tragedy, tragedy in which the opponent and the hero are indistinguishable.

With the progressive blurring of the oppositions between actants, actantial distinctions are at their weakest in irony. If Jake Barnes is the hero of *The Sun Also Rises*—and he certainly is in terms of the modal structure of the actants—then Robert Cohn would be the opponent. But the two are opposed less on the basis of virtue than of style. Jake is heroic because he knows how to behave. He adheres to Hemingway's code of "manly" stoicism. Cohn's sin is that he is a whiner who has the bad form to inflict his misery on others. Bill Gorton, Jake's sidekick, is not so much a foil to the hero as an equal. And Lady Brett, the "heroine," is hardly distinguishable from her masculine counterpart. She refers to herself as "chap," wears a man's hat, and adopts what Hemingway takes to be the masculine role of seducer. In this book that makes sexual difference function to designate actantial difference, such differences break down.

Toward an Actantial Typology

The position of the heroine in irony can help us to sum up the typology of narrative modes and genres in a final topology. In the Greimassian inscription of the ironic mode (see Figure 9), the heroine inhabits the fourth position, which Fredric Jameson describes as "decisive" insofar as it forces a rethinking of the category as a whole, in which all the elements are seen in "a different framework of meaning" (1972: 166; see Schleifer 1987a: 25–33). In each of the modes we have inscribed within the semiotic square, this position helps to define the related literary form, the related genre. In the heroic mode, it is inhabited by the helper (Figure 5), who can be conceived of as an aspect of the hero, what Greimas calls "the heroic nature of the hero" (1983b: 232). Such heroic power actantially defines melodrama, the literary form in which characters are so relationally defined and so formally defined (in terms of the receiver) that the helper implies the hero who is the receiver of the sought-for good. In the agonistic mode, the fourth position is inhabited by the "flawed" hero (Figure 7), who implies the opponent and is in a relationship of reciprocal presupposition with the helper (the receiver of the sought-for good). The position articulates, actantially, the flawed protagonist of tragedy. In the synecdochical mode it is inhabited by the opponent (Figure 8), who, in the purity of New Comedy, is a relative, usually a parent of the heroine who, finally, is the receiver of the sought-for good. As such, it creates the obstacles that are overcome in narrative and that define comedy. And finally, in irony the fourth position is inhabited by the heroine, who seems only a potential heroine, one about to be a sought-for good in a world where objects of desire are not clearly desirable. Irony is a literary form in which possibility overwhelms action and destroys narrative discourse.

Such, in any case, is a typology of actantial modes and genres, the systematic articulations of narratives conceived as cognitive structures. Most striking, we believe, is that Greimas's actantial categories do, in fact, suggest a cognitive level of analysis, between the level of contingent narrative events and purely logical or structural analyses of the semantic or cognitive content of the narratives. Narrative itself becomes, as Greimas says in *Structural Semantics*, "neither pure contiguity nor a logical implication" (1983b: 244). It is, instead, a vehicle for understanding that does not remain abstract but embodies, as we have suggested, cultural as well as cognitive values.

3

WHY ARE THERE OLD PEOPLE?

Narration, Natural History, and the Situation of Cognition

Semiotic Structures and Natural History

The narrative genres we examined in the previous chapter attempted to explore the *cognitive-semantic* structures governing the understanding of narrative. As such, they suggest more than simply literary genres. They suggest the cognitive logic of narration altogether. Still, as anthropologists note, all human cultures generate narratives, and it is legitimate to ask what functional ends such narratives serve in the economy of human life. In this chapter we explore this question in relation to the particular situation of aging members of human communities. To do so, we examine the adaptiveness of discursive cognitive behavior in general from the standpoint of the natural history of the species rather than the logic of understanding. Such a "natural history" is empirical rather than rationalist: it attempts to account for "facts" rather than to simplify phenomena, and in its descriptions of "objective" behavioral facts it does not matter, as Zellig Harris says of behavioral linguistics, "whether the system [of description] . . . is so devised as to have the least number of elements (e.g., phonemes), or the least number of statements about them, or the greatest over-all compactness, etc. These different formulations," he concludes, "differ not linguistically but logically" (1951: 9).

As we have seen in the previous chapter, the formal account of the cognitive semantics of genre is embodied in the receiver actant, which, because it is always syncretized with another actant in discourse, complicates the simplicity of binary logic. In this chapter, we emphasize the social aspect of cognition embodied in the sender actant, which, in its fundamental role as the speaker of messages, is multiplied in the *empirical* activities of speech. For Greimas, the sender actant is rather vague and undeveloped. As he said early in his career, his work originates in "a linguistics of perception and not of expression" (cited in Schleifer 1987a: xix)—which is to say, his work emphasizes the reception rather than the sending of discourse. For this reason, the sender remains *simple* in Greimas's analysis, simply the origin of cultural meaning, cultural power, and cultural goods. A focus on the sender can complicate our understanding of cognition from the vantage of the ecology of human, life just as Greimas's focus on the receiver complicates our understanding of cognition from the vantage of narrative semantics rather than (or along with) binary logic.

This is why we are focusing here on the cognition of old people. Such a focus leads us to an analysis of what Mary Hawkesworth calls "the conception of cognition as a human practice" (1989: 551). Instead of choosing the simple "component" as representative of "human cognition"—usually taken to be the mature adult in a hierarchy of human development, the high point on a linear curve in which the "sender" of fully developed cognition rises between the lack of childhood and the falling off of old age—a focus on aging requires that we conceive of the sender as *complex*. Such complexity is an element in what Evelyn Fox Keller calls an "organismic view" (1982: 125) not only of cellular or natural "order," but of cultural "order." Her opposition of law and order is particularly fruitful for our discussion. Law traffics in simplicities: simple data, simple logic, and a monumental or master narrative. But orders can be discerned that complicate the simplicities of law without descending into chaos or irrationalism or monological assertions of taste and intuition. One such order is the narrative order of natural history. Natural history is a narrative order precisely because it narrates the "events" of nature sequentially, with actors assuming roles as agents. The form of this narrative in our time is the Darwinian concept of adaptation articulated as natural selection. This concept entails what Stephen Jay Gould calls "a mixture of chance and necessity—chance at the level of variation, necessity in the working of selection" (1977: 12). "Natural selection," he says, "is a theory of *local* adaptation to changing environments" (1977: 45). It is a local rather than a monumental narrative.

In Greimas's account, the articulation of a logic of discourse creates a kind of monumental narrative. This is most clear in the actantial role he develops least, the role of sender. For Greimas, the sender is simply the subject of power over discursive meanings, discursive power, discursive goods. In a more complicated view, sending itself is a form of adaptation;

it is itself subject to narrative analysis. In this view, the sender is both the subject of power and the object of knowledge. In this chapter, by focusing on understanding the situation of old people within social structures—by examining an empirical-adaptive mode of explanation for the cognitive activity of the elder "component" of human social life—we try to give an example of Anglo-American empirical investigations of cognition while complicating the role of the sender of discourse. In Chapter 6 we return to old people as the subject of discursive power—the subject of rhetoric—to trace the cognitive-narrative strategies of the special case of their discourse.

The examination here, as in experimental psychological investigations more generally, is based on the assumptions governing Darwin's theory of natural selection. This theory focuses on the individual as the agent of the evolution of the species, and in so doing it takes, as Keller says, a single "component" of the species as the representative figure. In this chapter, we use this theory—with its attendant assumptions that simplify, generalize, and exhaust empirical data—to examine studies of cognition in old people. In the process, however, we articulate a description of aging that modifies the *positive* nature of natural selection by examining its *negative*—or, in terms of Greimassian logic, its "contradictory"—functioning. "Contradiction," Bruno Latour argues, "is neither a property of the mind, nor of the scientific method, but is a property of reading letters and signs inside new settings" (1986: 20). Here, we read empirically compiled data on senescent cognition within a setting that allows us to see a form of "preparedness" not in terms of positive attributes but in terms of the *least negative* attributes of aging, measuring that negation—"reading" it, as Latour says—in terms of social information. To do this, we examine empirical data on the cognitive behavior of old people collected by means of controlled experiments within a framework treating the elder as an information processor in biological and social contexts (see Abrahams et al. 1975 for a discussion of the difficulties and inconsistencies in defining "old"). In this setting, the data suggest that old people are uniquely suited—negatively as well as positively suited—to tell stories for other members of their social world. We argue, then, that they are uniquely suited to the actantial role of sender, especially in a context of the oral transmission of information.

Above all, this chapter attempts to demonstrate that seemingly simple facts—the object of Sellars's critique of the given—can be reconfigured to complicate perception and behavior in relation to cultural activities. Adaptation itself is one such fact, and Gould complicates it by emphasizing the *potential* rather than the *determined* nature of adaptation and by examining the *cultural* forms of "evolution" (1977: 251–59; 1981: chap. 7). The demonstration we are pursuing in this chapter is suggested by experimental data developed by cognitive psychology, data supporting the hypothesis that selective mechanisms exist making old people more effective in the articulation and oral transmission of certain kinds of information—of narrated knowledge—than younger tellers. Such "facts," however, exist within the framework of scientific and cognitive assumptions—the Darwinian framework of biological and cultural preparedness—which is summed up by the title of this chapter. The question of the title goes to the heart of an exhaustive accounting of cognition in which *measurable* cognitive changes that accompany aging can be understood (or at least questioned) in frameworks that have more explanatory force than the sad, accidental necessities of individual physiological degeneration (where "accident," like "cause" in other contexts, is an occultation of experience). This is why aging is such a useful object as an example of empiricism (as opposed to "ageless" literary forms we examined in relation to semiotic rationalism). With this focus we find that it is at least possible to frame questions about the adaptiveness of the "accidents" of physiological and cognitive aging and, more generally, about the narrative of adaptation. That is, the understanding of old people will help us to frame a broader sense of cognition and cognitive activity than either the unexamined empiricism of cognitive psychology or the logic of scientific semiotics describes.

Adaptation and Cognition

Darwin's evolutionary "explanation" of natural history—his suggestion that the "adaptiveness" of accidental changes in species serves to preserve those changes through "natural selection"—and the later extension of this explanation to include "preparedness" or the adaptiveness of the *propensity* for certain behaviors that can be subsequently learned—most notably cognitive behavior—are modes of explanation that focus most clearly on an *exhaustive* account of details in the world rather than simplifying their existence in terms of abstract principles and logic.

Of course, "natural selection" is a "simple" principle encompassing many of the assumptions we noted in describing psychological experiments in Chapter 1—the self-evident assumptions that an effect can be reduced to a cause, that simple behavior is easier and is more likely to be followed than complex behavior (conservation of energy), that differences and variety can be subsumed under general "generative" principles. But above all, natural selection, like natural history, is tied to empirical "details" of nature—including the overwhelming "detail" that nature is so full, that there are so many *things* in the world, that there need not be, in fact, the single "key" to its general pattern that Whitehead speaks of. Natural selection itself can, of course, be thought of as the "pattern of relationships . . . imposed . . . on external reality" that Whitehead describes (1967: 26), but it is a "pattern" whose results, like those described in contemporary "chaos" theory, are not of repeated figuration but of multiplying diversity. In other words, the natural selection of Darwinian evolutionism—the very principles of "adaptiveness" and "preparedness"—is a "theory" based on *accidents* rather than (logical) necessities, whose very motor is ongoing, complex, accidental change. This is the element of variation and change in Darwin that Gould mentions. By focusing (at least in part) on variation, Darwin offers the possibility that habitual, comfortable accommodation rather than simplifying apprehension might be the goal of cognition. Such "comfort" configures itself as a steady-state network of relationships that situates all members of a population (including old people) as alternatively central and peripheral to the functioning whole.

Within the framework of Darwinian explanation, it is appropriate to focus on human development and suggest that those characteristics of human development that are pervasive and persistent may be assumed to have been selected for their species-adaptive value. In addition to this "strong" form of selection, there exists a "weak" form of developmental or ontogenetic selection—that which is *contradictory* to the positive acquisition of traits or propensities to acquire traits—whereby those behaviors that are not weakened with age become predominant, *not* because of improvement in these abilities but because of the lessening of other behavioral activities. The binary opposition between strong and weak selection is a function, at least in part, of the social aspect of human life. This opposition, unlike the rational version of "contradiction" Latour describes, can be taken to be an empirically verifiable pattern in cultural life. Human social life literally creates the learned selection of behavior that Gould describes as "cultural evolution" (1981: 324). Darwin himself describes versions of this human or "artificial" selection as "man's power of accumulative selection" as an analogue to natural selection (1958: 48; see also Gould 1977: 41). Because living things, including people, are likely to repeat those behaviors for which they are rewarded, and since reward is more likely to follow the effectiveness rather than the ineffectiveness in serving society, we can expect old people to be more likely to perform those tasks for which their debilities are least in evidence. This "negative" selection of behaviors may occur because people receive the immediate benefits of their competence or because society is structured in such a way that certain behavior is rewarded symbolically. In either case, the situation constitutes a kind of weak preparedness which can help to identify the strong behavior that old people can rely on. Conversely, one would expect the weakest performance to occur in tasks that the aged seldom perform.

If the strong form of biological preparedness does exist in people, it should be possible to demonstrate that there is natural selection for longevity past child-rearing years in service to the social ends of the species. This is not the traditional view. Traditionally, the fact that many humans, unlike most species, live beyond their ability to reproduce has been most often regarded as selection for a long developmental and long parenting period. In most species, survivorship is inversely proportional to reproduction in that reproductive effort reduces survivorship and individual genetic fitness (see Wilson 1975). Thus, longevity past childbearing has been thought to occur because of variance in the natural selection process as humans are "selected," on average, to live at least as long as it takes to make their offspring self-supporting. T. B. L. Kirkwood and Robin Holliday even suggest that, since the germ plasm is potentially immortal, genes simply make use of a series of disposable organisms that, in effect, wear out after reproduction because of an accumulation of defects in bodies, or in macromolecules (1979: 97–99). This is a version of the "master molecule" model of genetics that McClintock argues against with a "steady state" model (Keller 1982: 124; Keller is citing D. I. Nannery). The same model is mocked in Samuel Butler's witticism that a chicken is simply an egg's method of making another egg. In an early version of this reductive notion, August Weismann (1891) seems unconsciously to parody Darwinian explanation in his contention that death itself is a form of "adaptation," occurring for the benefit of offspring in a process of natural selection so that trees, for instance, fall in forests in order that their young can pierce the canopy. These discussions of the "adaptiveness" of aging seem notably inadequate, even in light of their own grounding in natural selection. The first, discussing "inherent variance," is a mere description of phenomena claiming to be an explanation. The second is simply self-contradictory in saying that a "selection" for mortality benefits the species when immortality ideally would serve it much more economically in terms of the canons of generalizing simplicity we discussed in the Introduction. In this way, the agent of adaptation—the actantial role of sender—reduced and simplified to germ plasm, fails to explain the phenomena it examines.

Still, these traditional explanations of the adaptiveness of development have had important implications for the study of developmental cognition. Traditional views of cognitive change across the life span have implied a parabolic function in which infancy and old age are regarded as times of minimal competence, and "development" consists of adding or developing cognitive skills (i.e., "effectiveness") during the early part of the life span and subtracting it during the latter part. Recent work has tended to cast doubt on this inverted U model. Gisela Labouvie-Vief, for instance, reviews and offers a counterargument to this "linear" model, and much recent work in developmental psychology suggests a reciprocal relationship between individual growth and social contexts (see Labouvie-Vief 1982; Bowlby 1958; Brent 1978; Labouvie-Vief, 1977: 227–63, 1980). In many studies of early development, the young human organism is viewed as behaving effectively as it adapts to its environment, and there is increasing acceptance of the notion that infants and juveniles are biologically prepared to assume cognitive behavioral roles (Seligman 1970: 406–18; Sameroff 1972). Thus, the behavior of infants is currently viewed as instrumental to the success of the *species* (see Bell 1971). Gould even argues that neoteny—in which "juvenile stages of ancestors become the adult features of descendants" (1981: 333)—characterizes the flexibility of human development. In this argument, stages of developments are syncretized or superimposed.

The behavior of old people, however, has rarely been considered to have adaptive significance, nor are the behaviors of the old often considered in the light of their particular effectiveness. Even when empirical studies have argued and demonstrated stability in some cognitive skills among old people (Labouvie-Vief 1977: 227–63, 1982; Labouvie-Vief and Gonda 1976: 322–32; Schaie and Labouvie-Vief 1974: 305–20; Schaie and Strother 1968: 671–80), they have not suggested ways in which the *differences* between the young and old serve the species (see Brent 1978 for an exception). Recent reevaluations of the role of the old have not taken place in the context of explicit theories of social adaptation—within the context of the "organismic" view of biology Keller describes in McClintock's genetic work—and yet a focus on the cognitive functioning of the old can help us define more sharply the *situation* of cognitive activity within human cultures.

An "organismic" examination of cognitive activity as a form of species adaptation focuses on the relationship between cognition situated in one phase in the cycle of individual life—the phase of aging—and the larger human social context. It considers what Labouvie-Vief calls "the impact of socioecological conditions" on the individual (1977: 243) in terms of a larger "societal organism" Sandor Brent describes (1978: 23). Such an examination, like Zellig Harris's linguistics, focuses on the range of Whitehead's "details" rather than on the simplicity of Greimas's "logic." It makes the special case as important as the general case. As Labouvie-Vief has said, "any theory of adaptive competence must be willing to give up universalistic claims and instead define the social and historical context in which particular behaviors can be said to serve an adaptive function" (1980: 145). Behind our argument is the assumption that all evolutionary sociobiological theories, from Darwin and Spencer, work best—they explain the most—when they *multiply* questions about empirical details in the world, questions such as: "Why are there so many species?" "Why is there sexual reproduction?" and "Why are there old people?" An understanding of the adaptive functioning of human attributes of sexuality, of narrative discourse, of cognition in relation to human life in which the strong opposition between nature and culture (or biology and society) is no longer functional can be drawn from these questions. Specifically, we can explore ecological niches and sociocultural functioning of human cognition across the course of development from conception to senescence.

Recent suggestions argue for seeing aging as an adaptation in cultural as well as biological terms. Peter Mayer has reviewed a number of theories that argue for programmed aging (1979), but it was Sir Peter Medawar who first suggested that there may be a selection criterion for grandparents who survive to play a unique part in rearing the young (1952). Nevertheless, evidence that there is a selective behavioral advantage in any general population to include the aged was slow in coming, although A. D. Blest advanced that aging (and slow) saturniid moths are often caught by birds who subsequently "learn" that the whole species is unpalatable and, thereafter, refrain from preying upon young and reproductively potent moths (1963: 1183–86). In this instance the existence of the old enhances the overall (inclusive) fitness of the group, where inclusive fitness is defined as the effect upon the fitness—the survivability—of all the relatives multiplied by the degree to which they share genes with the old members (Dawkins 1974).

The effort to quantify empirically such understanding has also been undertaken in cognitive science. Robert Christian and George T. Baker have proposed a quantitative model that balances the cost of maintaining postreproductive, aged members with the benefits given by the aged to the group, their community (1979). In this model, the cost of maintaining the old members in terms of reduced resources is balanced with the benefits those members provide the group in terms of caring for offspring, serving as a distraction for predators, and strengthening social structure. The benefit gained from the aged members, when sufficient, outweighs the cost and, thus, increases the inclusive fitness of the group, the presence of old members outweighing the cost of maintenance. Using quantifiable data, Mayer has made convincing use of New England kinship data to support the positive adaptiveness of old people within a social group (1979). He points out that humans are the only species with menopause and that there is a significant postreproductive subpopulation in all recent societies. Most important, he argues that the sociocultural environment provides the opportunity for the natural selection of this subpopulation to function. In a social environment, old people increase the inclusive fitness of the group in which they live by providing their relatives with exposure to low levels of disease pathogens (thus conferring immunity), by choosing and preparing foods, by providing knowledge of scarce resources and dangerous substances, and so forth. Mayer compared the proportion of replica genes in the surviving blood relatives of a New England family and found that longevity was related to inclusive fitness. Those people who lived longer, in other words, had larger families in their reproductive years.

Like the standard logic of Darwinian conceptions of adaptation and preparedness, these investigations assume that the factors contributing to the survival of certain genes in a social group clearly may be at odds with the factors contributing to individual survival. The slow, old saturniid moth that Blest describes, for example, contributes to the survival of its kin precisely because of processes reducing its own prospects for survival. Similarly, "altruistic" behavior in birds (Dawkins 1974) involves a bird's sounding an alarm at the presence of a predator, even though such signaling puts the bird at risk. The genes of this bird, however, carried by its kin, will be more likely to survive, making the behavior of a "failing" older member adaptive for the group. In such explanations, again, is the urge to generalizing simplicity that does everything in its power to erase the special case in favor of generalization, even when, as in Darwin, the (generalized) individual is the focus of the generalization.

All of this work assumes the validity of simplifying *positive* science—data, numbers, empirical surveys—and does not emphasize the cultural complications of what Greimas describes as the contradictory or what we are calling, in the context of this chapter, "weak" preparedness. Still, the importance of such an evolutionary model for examining behavior in general, and cognitive activity in particular, is that it allows cognitive science to *situate* the universal existence of a *developed* skill—the activity of cognition—in the context of both the physical and complex sociocultural demands of a species or group within an environment. (For related discussions of the developmental-evolutionary perspective, see Bickhard 1979: 217–24; Birren 1980: 33–45; Ceci and Cornelius 1990: 198–201; Perlmutter, Kaplan, and Nyquist 1990: 185–97.) Human cultures at any one moment include individual members at various developmental stages, and these members are likely to contribute to (or detract from) the likelihood of survival of their group. To the extent that old people and others at different developmental phases are represented in the larger populations at diverse places and times, there is evidence that the presence of the aged increases the net inclusive fitness of their group, notwithstanding the drain on social resources they occasion and the personal tribulations of aging they endure (see Birren 1980; and Timiras 1972 for further discussions of the adaptiveness of old people).

In other words, the positive explanations of cognitive science suggest an organismic view in which we can multiply and complicate the adaptive advantages of old people within a social species. In such a view, we can situate the complexities of cognition—and the complex agents (or "senders") of cognition—within the contexts of culture, including the complex functioning of negation within cultural activities as well as narrative activities, in order to define more particularly the relationship between narrative and the institutions of understanding. The adaptive advantages of old people within a social species can be described in four ways:

1. As a result of the learning experiences they have had in the course of life, old people tend to perform some behaviors better than do younger people.

2. Old people have structural and physiological changes in the brain and elsewhere as a result of maturational processes. These facilitate some classes of useful behavior so that, for some behaviors, their performance will likely exceed that of other subpopulations in the group.

3. Old people show deficits in behavior as a result of their experience, such as learning habitual ways of behaving in various task situations that, in fact, impede effectiveness. Because learning is not monolithic but rather varies in amount or kind, old people are likely to be less deficient (e.g., less habitual) at some behaviors than others, and thus free other members of the group from some necessary behaviors (tasks).

4. Structural and physiological changes reduce the average old person's

ability to perform some tasks more than others—for example, they become less able to detect fine visual distinctions up close than at a distance—and these variations make them more likely to show better performance on those tasks in which change has taken less toll. In this way, they will free other members of the group for other tasks.

Thus, old people may "earn their living" by "strong" cultural preparation and biological preparedness (1 and 2), which make them more fit for certain tasks than younger people because new competencies emerge; or by "weak" preparation and preparedness (3 and 4), which tend to make them concentrate their efforts on tasks for which they show the least inability owing to age-related distrophy or experience. Furthermore, specific predictions about those behaviors at which old people excel can be made from empirical knowledge of physiological changes and of the kinds of experience that typically occur to people during the course of growing old (strong preparation). At the same time, logical inferences about the amount of relative decline in various abilities can be drawn from the frequencies with which these behaviors are manifest in the old and young respectively (weak preparation). Finally, it should be noted that both strong and weak preparation can occur because of the *social* experiences of the individual (1 and 3) or because of *biologically* determined alterations in the nervous systems (2 and 4).

Generalizing Cognitive Activity: The Elder as Information Processor

These four modes of understanding the biological and social adaption of age within the species can articulate a larger, more complex concept of "agedness" than the concepts derived from the accumulation of empirical "details" about old people or the simplifying logical development of a concept within *simple* positive and positivistic notions of logic and fact. That is, the very complexity of this conception is a function of "negative" as well as positive aspects of cognitive understanding. The concept of weak preparedness, like Darwin's concept of the positive adaptiveness of (accidentally) acquired traits, can focus and direct empirical research by articulating the complicating power of contradiction. On the semiotic square, this contradiction is articulated by the intersection of the levels of binary oppositions. In the four modes of describing aging we just outlined, "old" in its pejorative sense is described as both natural (physiological) degeneration (4) and cultural (learned) degeneration (3). This pejorative conception of aging, however, is contradicted and made complex by a positive conception of age, also articulated in these four modes of understanding agedness, which is the conception of "growth" (1 and 2). For this reason, these four elements of aging can be inscribed on a semiotic square.

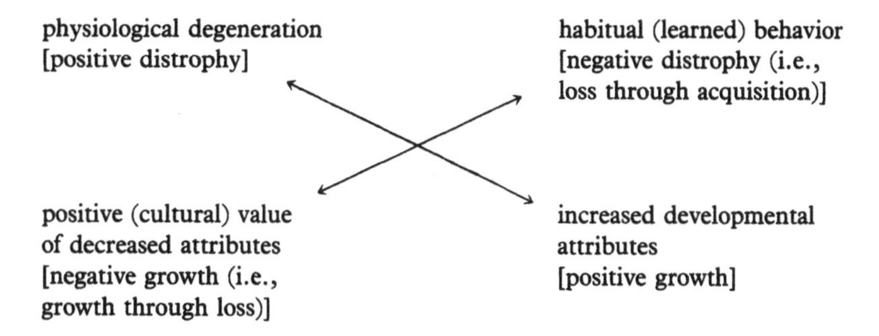

Within this complex conception of aging (understood within a cultural as well as a biological context), we can examine the cognitive activity of old people as members of a social as well as biological group. Here, then, we can survey and situate simple "empirical" data on the cognitive activity of old people that have been experimentally collected. The data describe old people whose physiology is changing and whose experiences have undoubtedly altered their ability to process information. Lenise Dolen and David Bearison have demonstrated the consistency between the processing demands and the processing abilities of the old (1982: 430–32), and we believe that this consistency between *cultural* demands and *biological* abilities exists because the binary opposition between culture and biology, like Lévi-Strauss's basic opposition between culture and nature, is too simple. That is, the cognitive contributions the old make to the societies in which they live, whether strong or weak, are defined as those that cannot be so well supplied by other members of their social group—a definition which makes the binary opposition between culture and biology locally irrelevant.

The elderly adult, by virtue of his or her developmental status—by virtue of the concept of "agedness" we have just presented in semiotic terms—serves the species by means of cognitive activity. That is, the elder is an experienced information processor embedded in a larger social context, and he or she functions, beyond positive and negative learned experiences, by means of physiological changes that alter information processing. The components of this cognitive processing system include the nature and amount of stimulation coming from social and nonsocial sources that is actually transmitted to the elderly; the integrity of physiological mechanisms that receive such stimulation; the attentional and cognitive systems involved in selection and interpretation of available sensory data (including the cognitive activity that compares data to memory content); and the decoding of information (including memory retrieval, habitual responses to stimuli, emotional responses to stimuli) that entails, most globally, whether or not cognitive activity is undertaken at all. This last factor, the undertaking of the cognitive activity of decoding altogether, is based on the judgment—conditioned by memory, habit, and emotion—that such cognitive activity will have an effect on the environment. This interrelated information-processing system of elderly adults—based, as it is, on the complex conception of the situated cognition of the aged—is poorly suited for rapid, interpretive encoding of information, but it is well suited for decoding and transmitting information already in this system. For this reason, old people are both strongly and weakly prepared (selected) to serve their cultures by means of cognition embodied in narration and discourse.

There is a good deal of empirical data to support the profile of an old person's cognitive ability as reduced but also specialized in these ways. For example, in some cultures that put a premium on speedy behavior (i.e., technologically advanced cultures) the elderly are required to retire, and the nature and amount of stimulation they receive tend to change radically. Also, in many cultures the appropriate social "approach" to an old person is formalized with etiquette and deference, ensuring that much information directed to the old is "presummarized" and delivered in prescribed form and often in a slow and ritualized manner. One experimental study found that young adults used a different language when asked to explain a task to an older as compared to a younger adult (Rubin and Brown 1975: 461–68). Another study showed that old people prefer greater physical distance from a middle-aged interviewer than do younger people (Feroleto and Gounard 1975: 57–61). These studies suggest that both the amount and quality of information given to old people are a function of age in both its cultural and biological dimensions.

There is also much evidence suggesting that the *quality* of information processing changes with age in ways that cannot be accounted for in purely physiological terms. By "processing," we mean more than the mere reception of stimuli but do not include any subsequent response. With this term, we refer to the very "cognitive activity" that cognitive science wrested away from the positivist behaviorism of stimulus/response governing psychology up to the 1950s. In Greimassian terms, the framework of information processing transforms a "receiver" into a "processor" if not a "sender." For instance, older people are more affected than young people by environmental conditions. Elderly comprehension of speech, for example, is severely reduced in a noisy room, even though, as John Corso notes (1977), old people generally show little loss in the recognition of speech frequencies. This effect in the cognitive mechanisms of old people is contrary to the expectation that the active central mechanisms will compensate for small reductions in sensitivity or for response limitations. Apparently, the opposite phenomenon occurs, and problems at the reception level of stimulation are further compounded by reduced ability at the level of cognitive processing. In the same way, some experiments (e.g., Botwinick 1978) present evidence of "stimulus trace persistence"—the persistence of stimulation beyond the point of providing for cognitive processing—interfering with cognition. This is stimulation that becomes general "perceptual noise" in old people (Layton 1975: 875–83) and creates problems with the processing rather than the reception of stimuli.

In effect, the aged often do not interpret rapidly changing stimuli as younger adults do and tend to perceive a steady-state environment even when there are discrete changes in stimulation. This perception has been measured by "fusion" experiments in which a row of lights is blinked on and off more and more rapidly, eventually creating the effect of solid light strings (Wolf and Schraffa 1964: 832–43). In another study, letters are projected on a screen and afterward "masked" by hatch marks (Walsh et al. 1979: 234–41). In yet another, when looking at ambiguous stimuli such as a hollow cube that presents a particular side as alternately a bottom or top, older people experience fewer perceptual shifts between the alternatives than younger people (Botwinick 1961). The data from these experiments suggest the existence of different modes of processing stimuli. Especially during divided attention tasks, as in trying to focus on one message in a noisy room, the elderly person has difficulty in blocking out the noise. Data from component selection and incidental learning tasks (Mergler et al. 1977)—tasks, for example, in which experimental subjects are asked to attend to the shape of objects and then are tested on an "incidental" characteristic such as the color of those objects—also suggest that the elder person processes less of all incoming information.

The study of the mechanisms of memory in old people is especially important in reconceiving the agent of cognition as a sender rather than a receiver. One particular focus on the "level" of information processing in memory highlights the organizational strategies used at the times of receiving and encoding stimuli (Craik and Lockhart 1972: 671–84). This approach asserts that stimuli encoded more elaborately or "deeply" at the moment of apprehension (e.g., the requirement to repeat a word's meaning and use it in a sentence rather than to number its letters) facilitate memory access and retrieval. Michael Eysenck (1974: 936) and Eileen Simon (1979: 115–24) found that the elderly have a "depth-of-processing deficit" so that "deep" encoding in long-term memory does not necessarily establish retrievable memory. They argue that the elderly cannot (or will not) use many resources at the point of encoding to elaborate or interpret stimuli that have not already been made evident by the experimenter.

Even more recent studies of automatic and controlled processing (Hasher and Zacks 1979), and of explicit and implicit memory (Hultsch, Masson, and Small 1991), argue that the elderly have less available capacity for effortful processing. On the other hand, Timothy Salthouse, Donald Kausler, and Scott Saults suggest "that connections must exist between age and cognitive performance independent of processing resources, and between processing resources and cognitive performance independent of age" (1988: 163). Regardless of the specific model of memory, however, considerable research data support the conclusion that the old do little spontaneous, elaborative processing of out-of-context verbal information, although they may still have the skills to do so (weak preparedness).

The question of what cognitive psychology calls the "storage" or permanency of information, once "in" the memory system, also bears on cognition. For example, in order to have stable (though not necessarily static) storage throughout adulthood, storage cannot be dependent on particular cells in discrete cerebral locations. Current theories of the brain mechanisms of memory that are part of the neuropsychological aspects of cognitive science we mentioned in Chapter 1 support this assertion. (See Gardner 1985: 262; Haecan and Albert 1978; Labouvie-Vief 1982; Strayer, Wickens, and Braune 1987.) Data on the availability of all potential memories cannot be collected independently of retrieval mechanisms, however, and extreme methods for gauging availability (by means of hypnosis and age regression, for example) have not been tested. One form of retrieval, research on reminiscence and life review processes usually conducted under the umbrella of "personality theory" (Butler 1963: 65–76; Erikson 1962), suggests that the elderly edit their specifically public reminiscences to enhance their own well-being in the present (Havighurst and Glasser 1972: 235–53; Romaniuk et al. 1978). Pleasant events, in short, become prominent while painful experiences recede. Whether memory is totally stored and only selectively retrieved or whether it is selectively stored is a moot issue, but empirical evidence suggests that the interpretation of semantic memory, which is often confused with retrieval, changes with age.

Nevertheless, retrieval of memories is a critical component of our understanding of the cognitive activity of old people. If, in fact, aging serves the biological-cultural ends of the species taken as an organismic whole, old people should show disproportionately high levels of recall for memories not in the experience of (younger) listeners. If aged people do indeed act as a repository of cultural lore and information, the critical material to be transmitted would be a remote memory for the elder, something that happened (or had been in its turn narrated) long before the younger listener was born. Even if there is no preferential access of very old memory material when compared to the access of old memories by younger people (see Botwinick and Storandt 1974: 303–8; Howes and Katz 1988: 142; Perlmutter, 1978: 330–45), old people are likely to remember many "old" events because proportionally more things have happened to them in the distant past (strong preparation). If the elder's current stimulus input is reduced, the proportion of older to recent memories becomes even larger.

Of course we are not talking about the commonly held notion of bad memory in old people—which even old people subscribe to (Loewen, Shaw, and Craik 1990: 43). Recent experiments show that the elderly remember salient, meaningful events from both their immediate and remote past. If the act of retrieval is viewed as a recoding event that will "strengthen" the existing memory (Schacter et al. 1978), the simple notion of "redundant" recall (which situates the elder as a passive "receiver" of memories) should be complicated to include information (i.e., salient events) that happened far enough in the past to have undergone considerable re-encoding. Furthermore, a generalized episodic or storylike mode that indicates the extensive elaboration of remembered information is often evident in these recollections (strong preparedness). Such narrative encoding functions to make the recalled information readily accessible for the listener in the same way that the structures of narrative genres make the series of narrative events a comprehendable "whole." It makes old people the senders of messages that do not simply or fully preexist them.

The empirical evidence points to a significant *difference* in information processing in the young and old, and, in the context of a study of the relationship between culture and cognition, we want to emphasize the complex adaptiveness of that situated difference. From this viewpoint, since the social environment no longer requires of the elderly (or rewards them for) rapid evaluation and processing of input which can be accomplished by younger members of the sociobiological group—Brent argues, in fact, that the elderly serve the group by their lack of speed and flexibility (1978: 30)—these changes in central processing of incoming stimuli are not unexpected. What *is* necessary behavior for the elderly involves interpretation—the decoding and transmission—of information. That is, the situation of old people (the biological and social situation) calls for cognitive activity that allows for the decoding and transmission of information that is already *in*—that, in fact, *constitutes*—the culture of which they are a part. This activity requires a stable, integrating memory system that can be readily accessed, yielding information in a semantic format; the ability to compare memory content to a single—probably symbolically coded—input; and proliferous output.

Recent evidence suggests that discursive and cognitive activities complicate many of the simple conclusions to which empirical data have led. It suggests that although the older person is not adept at many motor responses, he or she *is* well prepared to make verbal responses if no simultaneous encoding is required and if no constraints are placed on the style of response. One study found no age differences in reaction time to visual stimuli, if speech was the response instead of hand gestures (Nebes 1978: 884–89). In another study, Joseph Brinley reviewed a large number of experiments examining the relationship of the cognitive activity being tested and the instructions given to the subjects of these experiments (the "instructional sets"). This study concluded that anything that helps the aged to predict or anticipate stimuli does not facilitate their performance in the way that it does for younger experimental subjects (Brinley 1965). Focusing on a more explicitly discursive level of response readiness, Jack Botwinick (1978) advances that there is no evidence for greater behavioral rigidity with increasing age among adults in spite of the widespread notion that old people are inflexible. For example, Klaus Riegel and Roberta Riegel studied adult age differences for strength of agreement with a variety of attitudes (1960: 177–206). The elderly exhibited greater consistency in their attitudes than young adults with conservative attitudes only when propositions expressing the attitudes were stated in the third person. And Botwinick (1969) found that elderly adults tend to select a low-risk solution for a hypothetical career decision if one is available. But if risk could not be avoided, the elderly were as bold in their choices as younger adults.

A general concept that marks the discursive complication of empirical descriptions of cognitive activity is that of "effection," the process of constructing a response once a decision to respond has been made and then of providing feedback that a response has actually been made. An important part of information processing, effection is a concept that complicates simple stimulus-response models of behavior (the simple cause-and-effect model of empiricism) and simple sender-receiver models of communication (the simple expression-reception and diachronic-synchronic opposition of semiotics). Changes in this "feed forward" element of the cognitive processing of information can have extensive influence on the entire processing system in ways that suggest a steady state rather than a hierarchical order in cognition. Most important for the description we are constructing, weak preparedness can be understood in terms of effection. Weak preparedness suggests that the effective information processing of old people favors less effortful responses. Thus, effection accounts for the fact that old people are less likely to demonstrate something than to speak about it. In this regard, Patricia Cooper found no cognitive inabilities associated with aging when adults across a range of ages were asked to describe pictures orally. She writes "the age effect [is] not significant" for a picture description analysis (1990: 212).

Since effection may involve a "response trace persistence"—interference among a number of more or less relevant possible responses—it also accounts for the fact that the speech patterns of old people are more clearly broken into discrete response units than those of younger people—that in the speech patterns of old people there are longer pauses between idea units. Thus, Elizabeth Stine and Arthur Wingfield found older adults added more words and made more meaning-producing reconstructions than young adults when recalling short passages of speech (1987: 278). Other researchers found that older adults "produced more integrative or interpretive idea units" in story recall than did younger adults (Adams, Labouvie-Vief, Hobart, and Dorosz 1990: 24). The older adults seemed to create or superimpose a moral or fable—they seemed to add narrative structures—in the language they were memorizing. Such "response trace persistence" in effection also explains other changes in the coordination of response modalities that occur as a function of age. The combining of nonverbal responses—eye movements, facial gestures, and hand gestures—with verbal response may change or become more stereotyped as the effection system ages. Additionally, changes due to age in the reception of stimuli influence the amount of feedback the elder receives from his or her own responses. The old person may speak more loudly in the presence of others to hear his or her own speech. All these are matters for further testing, but they indicate the kinds of questions raised by the assumption of the adaptiveness of cognition in the elderly—the adaptiveness of *situated* cognition. In fact, from the standpoint of adaptation it is good to consider that the decreased frequency of effection (especially motor responses) and the associated decline in lean body mass in old people correlate with lowered food requirements (Karl et al. 1978).

Experimental Analysis of Oral Transmission

The seemingly simple data compiled by experimental psychology that we have surveyed can be used to support cultural and narrative conceptions of cognition if they are recontextualized in an examination of the interaction of old people with others in their social group. We have seen not only that old people are presented with certain classes of stimuli because they are old, but that, because they are old, they are listened to differently. In effect, old people form part of a social information processing chain that is "organismic" (steady-state). Both information given to and information expected from the aged are, at least in part, determined by their social situation and their cultural role. If the adaptive fitness of old people is indeed determined by the effectiveness of their sociocognitive skills of information processing and transmission, roles for the elder person ought to be *generalized* across cultures deploying the same aspects of information processing that characterize old people. Within a framework of Darwinian adaptation, it is unlikely that any group in society will consistently choose to take up roles for which they are unprepared. Here we are defining "role" as both social and physiological—in the same way the example of the cognitive activity of old people is simultaneously cultural and "human."

Some theorists emphasize the social aspect of this role and consider adaptation to aging as specific to the particular biological group or the particular community (Clausen 1972; Riley 1978). Other theorists, who argue for orderly developmental change in personality variables (Erikson 1962; Havinghurst 1968; Labouvie-Vief 1977; Schaie 1977–78), have emphasized a person's interpretation of his or her own adaptations. These particular views of aging and adaptation are quite unlike the simple empirical data of cognition we have cited or the generalized view of recent work in cognitive science that seeks to describe the adaptations of aging in terms of group fitness rather than individual experience. They are also very different from the more systematic examination of subjectivity and subjective cognition we will examine in the "special cases" of the psychoanalytic theories of Jacques Lacan in Part II of this book. Here again, as in the culture-specific examinations of the social role of biological aging—and as in Freudian ego-psychology which is the object of Lacan's fierce critique—what constitutes the individual subject of aging and the cognitive activities associated with aging is never examined or questioned.

The empirical data we have surveyed, which we have organized around ecological and social considerations of elderly adult cognitive behavior, suggest a generalized role of cognitive activity. This role, like an "actantial" role, is abstracted from the spheres of activity of old people. Those "spheres" are not derived from the logic of relations among activities but from the observation and measurement of cognitive activity. Still, as in all cognitive activity, the relationship between data and the assumptions and structures that direct observation is not simple. In the specific example we are pursuing, the generalized role of information processor—a complex version of the actantial role of sender assumed by old people—suggests that the elderly are likely to be more effective than younger people in transmitting cultural information to younger people when the message is consistent with the teller's role. Furthermore, the evidence we have, admittedly incomplete, suggests that the elderly will be more effective at oral transmissions than nonoral transmissions. Younger people will be more effective than the elderly in nonoral transmission of any information and more effective in the transmission of information not consistent with the sociocognitive role of the aged.

It is important to note that the kinds of preparation implied for this role of oral transmitter leave open (and thus complicate) the empirical question of whether the predicted differences result from biologically determined physiological changes with age (e.g., voice timbre, neural change regarding processing rate) or from increased speech effectiveness as a result of learning throughout a long lifetime. It may be that the changes in old people constitute only age markers to identify the age of the speaker for the listener (an identification that empirical testing suggests listeners consistently achieve) and that young listeners may come to treat speech from old speakers differently. However the mechanisms might operate, the notion that there would be such changes seems to follow from the model of situated cognition we are developing here and would be difficult to arrive at by means of a naive inspection of published experimental data or the elaboration of semiotic constraints. Once the differences are established—that is, once Latour's or Greimas's contradictions are noted—the mechanisms can be investigated. Before we discuss the empirical testing of this hypothesis more closely, we should examine the peculiar physiological nature of oral discourse as empirical and quantitative studies have described it. (For other, more or less "nonempirical" examinations of the difference between oral and written discourse—or, at least, examinations beyond the *measurements* of the following survey—see Derrida 1976, Havelock 1963, Ong 1982.)

Oral and written versions of language differ in the quality of both the message and the medium (Schallert et al. 1977). Oral transmission of information relies on the hearing of stress, intonation, and pauses and, as typically used, depends on extensive interaction of listener and speaker. Messages transmitted orally may lack the accuracy and precision of writing, but oral transmission also involves rich communicative cues (Dreiman 1962a, 1962b). The types of information carried by the medium of oral transmission can be studied even when the interaction of listener and teller is eliminated. Listening to tape recordings of voices does not allow for any visual cues, nor does it reflect the dynamic interaction of speaker and listener. It is a kind of degraded oral transmission. However, the listener's implicit knowledge of the speaker in such a situation is quite astounding. The age and sex of the speaker (Mergler et al. 1985; Ryan and Capadona 1978: 98–102; Shipp and Hollien 1969) and his or her height and weight (Lass and Davis 1976; Lass et al. 1979) can be reliably estimated even when the listener is not cued to answer these questions until after the oral transmission.

Determination of the age and sex of a speaking voice may depend on distinct qualities of the voice being judged. Particular vocal qualities that change with age have been assessed for both male voices (Mysak 1959; Ryan 1972) and female voices (Charlip 1968). Edward Mysak found that pitch levels of old male voices were higher than in middle age and exhibited greater variability in voice frequency (1959: 46–54). William Ryan recorded male speakers who had normal hearing and found that the vocal intensity increased and the overall rate of speaking decreased with age in recorded samples of oral reading and in impromptu speaking (1972: 265–68), and more recently Julie Liss, Gary Weismer, and John Rosenbeck found some differences between acoustic characteristics of speech production in young and very old men (1990: 35–36). The concept of "normal" hearing, however, may imply a sampling bias against "normal" aging individuals and in favor of the single "component" of mature adulthood, since most aged individuals show hearing loss. The data from Walter Charlip's (1968) analysis of female voices were less definitive. Recordings from females between the ages of forty and ninety with normal hearing revealed no age change in average pitch, variability of vocal frequency, or any other factor considered. Yet the age of the people with those voices could be determined. Charlip suggests that the problem of what makes one voice sound older than another may involve complex factors such as the timing and accuracy of muscular action that influences articulation of consonant-vowel combinations.

On the basis of this empirical (and predominantly quantitative) work on oral discourse as well as the more general conclusions about the cultural situation of old people that we examined earlier in this chapter, one two-part experiment predicted that old people would be more effective than younger people in oral story presentation (Mergler et al. 1985). The experiment analyzed the assessment and recall by young adults of vocal recordings of young, middle-aged, and elderly speakers of both narrative and descriptive prose passages. More positive listener assessments were elicited when young voices transmitted the descriptive passage than when older voices relayed the same information. When narrative prose was narrated by older people, however, listeners recalled more material accurately. In other words, the assessment of the oral message in this study was also affected by the semantic content and structure of the message.

While some questions remain about the exact physical qualities that distinguish an older from a younger voice, it is clear that physical cues are available for age grading of oral information. If these age cues are contradicted by specific information in the oral message or by additional third-party information, it is not clear how evaluation and recall of the message could be changed. In the second part of the experiment, young adults listened to recordings of middle-aged voices reading a narrative prose passage. These listeners were told that the voices were of young, middle-aged, or old people. Overwhelmingly, subjects acted on the false information about the age they were given. When thinking that the voice was that of a sixty-five-year-old adult, the young adults generated three times as many unique comments regarding the speaker, including both positive comments ("wise," "distinguished") and negative comments ("cunning"). Yet, recall of the prose material did not show the age-of-speaker effect of the first part of the experiment in relation to the false ages. Thus, recall seems to be affected by physical qualities of the medium that vary systematically with real age but not with "manipulated" age.

It is just this kind of experiment, examining one particular sociocognitive function of the elder, the actantial role of sender, that present knowledge of the cognitive transmission of old people suggests when this knowledge is examined in the context of the possible adaptive function of aging for the species as a whole. If cognition in old age results from a form of preparedness, there will be physical or psychological change leading to greater inclusive fitness. This last experiment suggests that vocal changes or other characteristics of the recorded story, as yet unquantifiable, allow more effective information transmission leading to better recall of the message. This, in turn, may lead to an age-specific, sociobiological role of sender which is compatible with this age-specific skill. If the type of message being transmitted is compatible with other age-specific stereotypes, the evaluation of the message is more positive. In this case, highly structured prose with an explicit moral—the discourse of narrative cognition—resulted in a measurably greater amount of transmitted information when it came from an old person. That is, listeners found (and demonstrated in response to test questions) that old people were the most effective "senders" of narrative prose with an explicit moral. In Chapter 6, we analyze the cognitive structures of this discourse. Here what is most important is the ways in which simple empirical testing of cognitive development can—and must—be complicated with consideration of the steady state of social life and the cultural situation such testing encompasses.

Old people are complex senders of cultural information encoded in narrative form. That is, old people make themselves, or find themselves to be, the "senders" of narrative discourse (see Nussbaum, Thompson, and Robinson 1989; Obler 1980). As a grandfather of ninety, Oscar Schleifer, told his grandson, "I enjoy talking with you because, when you get to be ninety, you need all the reasons you can find to get up every morning." This explanation combines strong preparedness—the "knowledge" of life the old man had to impart—and weak preparedness—his "thoughtfulness" occasioned by the enforced idleness of physical decay and the fact that the decay itself gave him a different, a "thoughtful" rather than active, way of dealing with the world. Together, the strong preparedness of experience and the weak preparedness of finding discourse, rather than meaningful activity, the only way of discovering the significance of experience tend toward narrative discourse. They do so the same way narrative genres enact the relationship between a more or less anonymous "sender" and the complicated "receiver" of literary narrative. In the sociocognitive role of the elder, cognition itself—the cultural activity of cognition—finds voice by finding, in all its complexity, its aging sender.

Natural History and Narration

While the data of these experiments are admittedly tentative, they are presented here as much to exemplify as to instantiate empirically the kinds of *testable* (i.e., measurable) questions that cognitive psychology pursues. They can suggest this double goal because they arise from the *complex* assumption that cognition itself can be situated within a framework of sociobiological adaptation. Such generalization, as Greimas says of narrative, is "neither pure contiguity nor a logical implication" (1983b: 244)—neither purely empirical nor rigorously logical. Nevertheless, this generalization (like the actantial generalizations discussed in Chapter 2) suggests a certain logic of aging. Highly structured prose with an explicit moral is a form that makes experience in time meaningful. Like narrative genres, prose with a plot and an explicit moral presents the experience of interrelated positions—the interrelationship of the actants of discourse that produces narrative plot—and the cognitive closure of that plot, marked in Greimas's analysis of narrative discourse with the syncretism of a receiver with an actant. We have also plotted natural history in the context of the cultural studies of this book and have offered the cognitive closure of our plot by means of the syncretism of a sender with a person at a relationally defined stage in human development.

Jerome Bruner has argued that "well-formed narrative . . . [is] capable of combining in a single structure the canonical and the aberrant, the normative and the descriptive, the internal subjective and the outer objective" (1990: 349). Recently, Greg Myers has maintained that natural history itself depends on narrative structures that aim at creating the "sense of an immediate encounter with nature" (1990: 194). "Natural history," he writes (paraphrasing E. O. Wilson's *Sociobiology: The New Synthesis*), "is the opposite pole from developed theory" in that it presents "a written account of actions of particular animals at a particular place or time, recorded by particular observers" (1990: 195). Such narrative texts, we have argued (following Greimas) narrate neither the pure contiguity of accidental events nor the logical implications of theory. Instead, like Darwinian adaptation, they mix chance and necessity: the "chance" of various "special cases" and the necessity of a "moral" to the story. In natural history, as Myers says, "things happen. Such events are indicated in natural history texts . . . by the use of the past tense. In contrast, the present tense usually indicates, in scientific texts, the general nature of the phenomenon being described, asserting that it is true at all times" (1990: 196). Besides the past tense, natural history offers apparently gratuitous details, the treatment of animals as individuals who are often "like characters in novels," and the characterization of the human observers of "nature" (1990: 196–201). Above all, "natural history texts seek out the singular, whereas biology texts seek out the typical" (1990: 204).

Myers's textual description of "natural history" also describes the narrative structure of the argument concerning aging and the aged sender we have constructed from the attempts of empirical cognitive psychology to isolate and present the "general nature" of cognition in humans generally and in old people specifically as a class of humans. In this chapter we have arranged empirical evidence concerning cognition—evidence that often measures isolated cognitive behavior within severe physical and conceptual limits—to tell a story of sociobiological activity in which the elder is the sender (as in Propp and Greimas), but in which the activity of sending is complicated to include modes of processing and "effection" that make the elder *part of* the cultural meaning, cultural power, and cultural goods that, as sender, he or she presides over. In other words, the story we have told is organismic rather than hierarchic, *cultural* in a wider sense in which it somehow encompasses society and biology, the differing versions of "culture" presented by Matthew Arnold and E. B. Tylor. Such an account, like the power of the narratives of natural history Myers describes, has the ability to provide comfort with its form of cognition. That is, like a story heard over and over again, it arranges events and still leaves room for all the participants who, in the steady state of narrative balance, rotate through complex situations. Like the eighty-four-year-old speaker we encounter in Chapter 6 who says, "I tend to look upon other old men as *old men*—and not include myself . . . because it is still natural for me to be young in some respects" (Blythe 1979: 185), the "natural history" of aging is a complex history whose actors never simply play one role.

This complexity of roles suggests that models of adaptation, mixing as they do chance and necessity, are particularly suited to the narrative forms of natural history. Just as the literary genre of irony is characterized by potential rather than fully realized cognitive information, the narrative of natural history assumes a similar form. This form, like the other forms of narrative genres, is characterized by complexity and complication. By crossing natural history with cultural history, the study of culture and cognition allows us to integrate biological and cultural evolution and to situate cognitive activity as simultaneously cultural and natural. The extreme of a simple cultural view has been stated by Theodosius Dobzhansky. "Culture," he writes, "is not inherited through genes, it is acquired by learning from other human beings. In a sense, human genes have surrendered their primacy in human evolution to an entirely new, nonbiological or superorganic agent, culture" (cited in Wilson 1975: 550). The extreme of a simple biological view has been stated by E. O. Wilson himself. "The transition from purely phenomenological to fundamental theory in sociology," he writes, "must await a full, neronal explanation of the human brain. Only when the machinery can be torn down on paper at the level of the cell and put together again will the properties of emotion and ethical judgment become clear. . . . Stress will be evaluated in terms of neronal perturbations and their relaxation times. Cognition will be translated into circuitry" (1975: 575).

The complex superimposition of these two approaches, like the superimposition of narrative genres found in ironic texts, can be seen in what Gould calls "biological potentiality" as opposed to biological determinism. Such a "potential" reading of adaptation traffics in negativities without the binary opposition of nurture and nature (1981: 320, 330). Instead, adaptation can be seen and narrated in its "flexibility." The "markedly increased brain size in human evolution," Gould writes, "added enough neural connections to convert an inflexible and rather rigidly programmed device into a labile organ, endowed with sufficient logic and memory [the combination of logical implication and pure contiguity] to substitute non-programmed learning for direct specification as the ground of social behavior" (1977: 257). Gould calls such "non-programmed learning" an "adaptive, but nongenetic, cultural trait" (1977: 256). In *The Mismeasure of Man* he adds that "even when an adaptive behavior is nongenetic, biological analogy may be useful in interpreting its meaning. Adaptive constraints are often strong, and some functions may have to proceed in a certain way whether their underlying impetus be learning or genetic programming" (1981: 327).

Like the cognition of aging we have discussed, neuroscience, in Howard Gardner's description, also presents such complex models. Thus, Eric Kandel argues that "the potentialities for many behaviors of which an organism is capable are built into the basic scaffolding of the brain and are to that extent under genetic and developmental control. Environmental factors and learning bring out these latent capabilities" (cited by Gardner 1985: 280). But most interesting in the context of this chapter and of our wider discussions of narrative throughout this book is the analogy between neurocognition and holographic recordings developed most fully by Karl Pribram. "A hologram," Gardner writes of Pribram's work, "is the plate or film with the recorded pattern: information about any point in the original image is distributed throughout the hologram, thus making it resistant to damage. Since waves from all parts of the object are recorded on all parts of the hologram, any part of the hologram (however small) can be used to reproduced the entire image" (1985: 283). In this model of neurological activity, in the same manner that "many holograms can be superimposed upon one another, so can infinite images be stacked inside our brains" (1985: 283). This superimposition repeats the flexibility by which Gould characterizes human biological and cultural adaptation and takes its place with the superimposed networking of science that Latour describes. Moreover, it takes its place in the superimpositions of the actantial roles and modes of narrative we described in the preceding chapter. Finally, like the old man of eighty-four we quoted a moment ago—for whom it is still natural to be young in some respects—it models a steady state of development in which different stages all contribute to human cognitive activity and any particular role is always complex. This complexity calls for narratives and negativities—neither pure contiguity nor logical implication—such as the "natural history" of aging we have woven from recent work in cognitive psychology in our attempt to answer the cultural-biological question "Why are there old people?" in terms of a natural history of cognition.

Part II

CASES OF COGNITION

4

SPECIAL CASES

Freud, Einstein, and the Dream of Understanding

Overdetermined Narratives

In this chapter we examine particular cases of cognitive activity in relation to the general (and generalizing) case of narrative cognition. Specifically, we examine the implicit theory of narrative in Jacques Lacan's reinterpretation of Freudian psychoanalysis as a version of the semiotic logic of cognition we discussed in Chapter 2. We also examine the explicit theory of narrative in Bruno Latour's interpretation of the theories of special and general relativity as a version of the cognitive generalization implicit in the empiricism of scientific knowledge. In both cases we focus on the cultural elements of cognition. If cognitive science, as we noted, assumes in the scientific methodology of its empirical studies that an effect can be reduced to a cause, that simple behavior is easier and is more likely to be followed than complex behavior (conservation of energy), that differences and variety can be subsumed under general "generative" principles, then Freudian psychoanalysis complicates these assumptions even while it participates in them.

It does so, in large part, by superimposing cultural and psychological phenomena. This is what Freud calls the "overdetermination" of psychological phenomena. Psychological life—including, as we are arguing throughout this book, cognition—is irreducibly complex, and in psychoanalysis that complexity takes the form of the *overdetermination* of effects (the multiplication of "causes"), the inherent *nonsimplicity* of behavior (the inextricable complex of physiological and semiotic aspects of human existence), and the *nonhierarchical* relationship between special and general cases (in Freud, the dissolution of the distinction between neurotic and normal behavior). These complexities of psychological life in psychoanalysis find their articulation in the method and focus on narrative in Freud's work—his "case history" approach to psychology and his "talking cure" therapy for neurosis. The focus on narrative is not always explicit in Freud. Like Greimas and Courtés, he would distinguish between method and epistemology precisely because, like them, he eschews traditional philosophizing in favor of "scientific" rigor.

Jacques Lacan's "return" to Freud focuses on the semiotic (if not the narrative) aspects of psychoanalysis, and in the early 1980s literary critics frequently viewed Lacan's idea of discourse as the basis for a theory of narrative—especially Jean Bellamin-Noël, Rosalind Coward and John Ellis, Barbara Johnson, and others (see, for instance, the contributors to Davis 1984). At the same time, the radical dimension of Lacanian theory was evident in film studies and French feminism, as evidenced in the work of *Screen* magazine, Christian Metz, Stephen Heath, Luce Irigaray, Michelle Montrelay, and others. All of this work, like that of Lacan himself, participated in the questioning of the function and unity of the psychological subject—the unity of "mind"—initiated by structuralism and poststructuralism in order to articulate a model of textuality and, implicitly, of cognition. In this questioning, the unity of the experimental subjects of cognitive psychology we examined in the preceding chapter was called into question in the same way the assumptions governing those experiments came to be seen as "overdetermined." These critics, despite the wide range of their fields, assumed that the subject of cognition was structured like a language and, more specifically, like a narrative, and that as with language its apparent unity was marked, and indeed enabled, by the (complex) disruption of that unity. In language, that disruption is the lack of congruence and compatibility between the significatory and communicative functions of discourse. This incongruity participates in the larger incompatibility in cognitive activity between the *constructed* nature of knowledge and its apparent self-evident truth, the phenomenalistic and physicalistic conceptual schemes Quine describes (1961: 17); it participates in the failure of the harmony of reason and the world that Whitehead assumes and the breach between method and epistemology Greimas assumes. Whether linguistic structures condition and govern cognition or whether cognitive structures determine the nature of language is irrelevant here, even if this opposition—another binary opposition—marks the global difference between Anglo-American cognitive science and Continental semiotics.

In either case, the critics of narrative and discourse we have mentioned, working within the Continental tradition of semiotics, reasoned that in formulating the semiotic connections between psychoanalysis and language, and in defining the subject of psychoanalysis in semiotic terms, Lacan contributed a description of how narration works—especially in relation to plot or "experience" (the manifest text)—to produce understanding. Lacanian modeling of the Freudian subject along the lines of the cognitive-narrative structures governing discourse, in short, provided a basis for the reconceiving of classical narrative tropes, just as cognitive science reconceived the adaptive functioning of discourse by modeling individual activities of discourse and cognition in relation to Darwinian conceptions of the relationship between individual and species. In the Lacanian model of narration, metaphor, metonymy, deferral, and displacement all have their counterparts in ancient Greek rhetoric and poetics (which themselves are early models of cognition) but are now recast as fundamental psychoanalytic processes positioned in the narrative of a psychological "working through."

Central to this conceptual framework is the "subject," not in the classical mode of an autonomous subject of perception and "experience"—based on the model of the Western (and even Darwinian) "great individual"—but the "subject" conceived as a linguistic and rhetorical effect, a synthetic creation analogous to perspective in representational painting. In this transformation of the role of "subject," "mind" is conceived not as a substance or thing but as a substantified "effect"—a set of overdetermined relationships that create the phenomenal effect of a thing to which one can attribute qualities. Thus, narration is a playing out, a theatricalizing of the Oedipal conditions necessary for the subjective "effect" to take place. By "Oedipal conditions"—clearly a charged (or "overdetermined") Freudian metaphor that often generates great controversy—we simply mean the unavoidable fact that human beings are born into family and social structures. As Lévi-Strauss and Lacan suggest, we are born into a network of linguistic and cognitive structures that condition "experience" even when experience seems immediate and knowledge self-evident—so that our basic sense of subjectivity, subjective experience, and true knowledge are historically determined "effects" rather than essential and transcendental aspects of ourselves and our world.

In describing Lacan's narratology, however, recent critics have said little about it in relation to larger cultural commitments and ideology, the relation of culture and cognition. If, as we argue here and in the next chapter, Lacan's narrative schemes build upon and develop a complex sense of cognition, then cognitive activity—the cognition of language and adaptation we have already examined—should be understood in relation to culture. In other words, the study of cognition, most notably in the radical questioning of the function and unity of the subject of cognition in Lacan (but also in Darwin), lends itself to questioning and opposes seemingly self-evident and transcendental "truths" about culture, human nature, and knowledge. Despite this logic, however, the inherent articulation of social contentions within psychoanalytic discourse, principally its reliance on the conflict of "subject" and "other," has not had an especially strong impact in literary studies. It has had little impact on cognitive science—even though Lacan's situating of his work in *social* relations of opposition, as we will show in reference to Freud's dream of "Irma's injection," points up the potential of his discourse as a way of appraising the ideological investments in cognitive as well as literary and cultural studies. Lacan situates his thinking in explicitly "oppositional" terms, the traditional theoretical frame and discourse of ideology, and we propose to examine, within his psychoanalytic framework of understanding, the cultural implications of his description of cognition. In this examination we attempt to take his reinterpretation of Freud, as he seems to insist, as a form of overdetermined oppositional critique that presents the special pleading of ideology and politics as well as the generalizations of cognitive "knowledge."

Lacanian narrative theory clearly marks a specific management of power in the world, and the neglect of this issue is a failure to place Lacan (and other descriptions of cognitive activity) in a sociocultural context. His is a practice, moreover, aligned with other attempts at oppositional cultural criticism, and the tradition of oppositionalism runs equally through Lacan's thinking and points up its implications for ideology. Such implications, we think, can be discerned in less controversial cognitive (and "scientific") activity than that of Freud or Lacan, and for this reason we examine the cultural significance of cognitive activity in the example of Latour's sociological understanding of Einstein as well as Lacan's semiotic understanding of Freud to show that Lacanian narratology itself is not a "special case." In both, the generalizations for cognition suggested in Greimas's analysis of narrative will be apparent.

The Dream of Dreams

The functioning of Lacanian narratology in relation to cultural cognition is evident in Lacan's analysis of Freud's dream of "Irma's injection" in

The Interpretation of Dreams, which Lacan calls the Freudian "dream of dreams" (1991: 147; see also Weber 1987: 73–85 for an analysis of Freudian dream analysis). The Irma text is especially revealing for its exposure and treatment of a psychoanalytic will to power, a political dimension of this dream that Freud notes and marks but simultaneously avoids. In fact—aware of disturbing implications for psychoanalysis in this specimen dream—Freud raises significant issues of semiotic dream interpretation and then subverts them by analyzing the dream in evasive and personal terms. Here is Freud's narration of the dream of Irma from *The Interpretation of Dreams:*

> Dream of July 23rd–24th, 1895
>
> A large hall—numerous guests, whom we were receiving. —Among them was Irma. I at once took her on one side, as though to answer her letter and to reproach her for not having accepted my "solution" yet. I said to her: "If you still get pains, it's really only your fault." She replied: "If you only knew what pains I've got now in my throat and stomach and abdomen—it's choking me"—I was alarmed and looked at her. She looked pale and puffy. I thought to myself that after all I must be missing some organic trouble. I took her to the window and looked down her throat, and she showed signs of recalcitrance, like women with artificial dentures. I thought to myself that there was really no need for her to do that. —She then opened her mouth properly and on the right I found a big white patch; at another place I saw extensive whitish grey scabs upon some remarkable curly structures which were evidently modelled on the turbinal bones of the nose. —I at once called in Dr. M., and he repeated the examination and confirmed it. . . . Dr. M. looked quite different from usual; he was very pale, he walked with a limp and his chin was clean-shaven. . . . My friend Otto was now standing beside her as well, and my friend Leopold was percussing her through her bodice and saying: "She has a dull area low down on the left." He also indicated that a portion of the skin on the left shoulder was infiltrated. (I noticed this, just as he did, in spite of her dress.) . . . M. said: "There's no doubt it's an infection, but no matter; dysentery will supervene and the toxin will be eliminated." . . . We were directly aware, too, of the origin of her infection. Not long before, when she was feeling unwell, my friend Otto had given her an injection of a preparation of propyl, propyls . . . propionic acid . . . trimethylamin (and I saw before me the formula for this printed in heavy type). . . . Injections of that sort ought not to be made so thoughtlessly. . . . And probably the syringe had not been clean. (1965: 139–40)

In this narrative Freud presents the dream in four increasingly pointed indictments of his own relationship with Irma. First, he tells how he and Irma meet at a party and accuse each other of bad faith in their therapy sessions. She only suffers, he claims, because she rejects his "solution" (left unstated) to her problem. To this she says that he refuses to recognize the pain in her "throat and abdomen" and that she is "choking" all the while. Second, Freud then becomes "alarmed" at her "pale and puffy" appearance and admits that he could be wrong about her condition. He pulls her to the side of the party for reexamination, which at first she resists in the manner of "women with artificial dentures." At last, peering into her mouth, he finds a "white patch" on the side of her throat and "whitish grey scabs" on "curly structures" attached to the sides of her throat. Increasingly worried, in the third moment of this account, Freud calls in "Dr. M"—who was also "very pale," "walked with a limp," and was "clean-shaven"—for additional diagnosis. Next appear "my friend Otto" and "my friend Leopold" also for more diagnosis. Leopold then takes over and examines Irma chastely—without removing her clothes—and discovers an "infiltration," an infection, on her lower left abdomen and another one on her "left shoulder." Next, Dr. M announces—mistakenly, as Freud later comments—that "dysentery will supervene" anyway to solve this problem, and the "toxin [of the infection] will be eliminated" naturally and automatically through her bowels. In the fourth and final section of the narrative, Freud turns to Otto and offhandedly indicts him as the cause of the infection, for Otto, it turns out, previously injected Irma with "trimethylamin"—a "sexual" fluid "not to be made so thoughtlessly"—while using a "syringe" that "probably . . . had not been clean."

In his interpretation, methodically dividing the dream into segments, or separate signifying functions, Freud emphasizes three semiotic, or structural, tenets of interpretation—suppositions, as he says, for approaching the "characters and syntactic laws" of the dream text (1965: 312). The first is that the dream text is composed not of a "natural" language in the sense of being inherently meaningful but of signifiers with referents and meanings yet to be established. The dream, in other words, is a cognitive structure that *must be interpreted*, and the referents for signifiers in this dream will be determined precisely by the narrative activity specific to this text. Second is that segments of the dream, or signifiers, are fixed in a sequence or order that limits their interaction within the dream and stabilizes the set of referents (at least in broad terms) the dream will evoke. Freud, like Propp in his very different analysis, makes narrative sequence—the relatively fixed order of narrative—a distinctive feature of narrative cognition. Semiotic interaction (semiosis), it follows, will occur loosely within this order. Third is that other syntactic and symbolic restraints on the dream text combine to produce a center, or structural intention, to the dream—implying (substantifying) a kind of dream "subject" for this intention. Viewing the text as a whole as signifying this intention, Freud says that "*when the work of* [dream] *interpretation has been completed, we perceive that a dream is the fulfilment of a* [single] *wish*" (1965: 154). He is arguing, in effect, that as an organized textual system—a system of cognition—the dream narrative evokes a functional pattern that reflects its operation.

Thus, Freud's analysis of the dream methodically emphasizes these semiotic, or structural, tenets of interpretation and cognition: the differential nature of the sign, its inherent *binarity;* semiotic interaction (or semiosis) within the text; and structural intention, or a "subject," generated in the act of performing—or interpreting—the text. Drawing from these tenets, he focuses on the "contamination" pattern in this dream, specifically that Irma has been violated by the inappropriate "injection" of "sexual" fluid into her body. Leopold and Otto represent behavior toward Irma that, in fact, belongs to Freud as well as to his reversed double, the incompetent "Dr. M." That is, whereas Leopold, a "good" doctor, approaches Irma chastely and, in so doing, advances her treatment with an insightful and efficient examination, Otto, the inept doctor, violates standard medical precepts by relating to Irma sexually and, as a result, makes her more ill. This emphasis on sexual and therapeutic "contamination" explains Irma's initial symptoms and accounts for the "infiltration" (infection) of her shoulder and abdomen. Finally, the contamination theme also arises in Freud's commentary as he accounts for the various details of Irma's condition—drawn from his own professional and personal experiences—that comprise the dream narrative.

But while Freud grounds a semiotic methodology in this "dream of dreams"—which is, in fact, an important early manifesto of the semiotics of Continental cognitive theory—he concludes his commentary with purely personal, nonsemiotic ("empirical") referents. In his commentary, for instance, he explains dreaming about misdiagnosis by referring to his prior anxiety two years before over the "serious [nasal] illness" of his daughter (1965: 143). During that period he also encouraged cocaine usage to patients who later suffered under the drug—to the extent, in one case, that "the misuse of that drug had hastened the death of a dear friend of mine" (1965: 144). These biographical referents identify the dream as masking a supposed "actual" struggle Freud had over bad advice he gave. He eventually closes off interpretation by saying that "certain other themes played a part in the dream, . . . but when I came to consider all of these, they could all be collected into a single group of ideas and labelled, as it were, 'concern about my own and other people's health—professional conscientiousness'" (1965: 153). This interpretation is exceptional for its idealism, its lack of irony or ambiguity, and its decidedly non-Freudian avoidance of ambivalence and the "oppositionalism" of semiotic analysis. In it, the phenomena to be interpreted preexist their understanding and are substituted for the dream text, just as the object of cognition—the referent—preexists the knowing subject in the classical paradigm.

Jeffrey Mehlman has argued that in reading this dream as he does, in terms of such personal referents, Freud actually resists the cognitive textual implications of psychoanalysis—surely its narrative force—and is "reducing dreams to the stability of a fixed meaning assimilable to the ego" (1981: 180). That is, Freud's underlying claim for this dream, contrary to the stated aims of psychoanalytic dream interpretation, is that semiotics has uncovered the nonrepresentational, preexisting, or "essential" origin for the dream (the dream thought and wish fulfillment) in his own experience—his daughter's illness, his urging of cocaine use, and the death of his friend. However, in that Freud reads the dream as a camouflaged representation of an already constituted experience, already "known" to him—like a picture with a familiar figure hidden in it (rather than the structural ambiguity of an optical illusion we present in the next chapter)—the apparatus of dream interpretation in this instance, as Mehlman argues, "is but the subtlest ruse of narcissism" (1981: 181) and has no bearing on the narrative dimension of (dream) texts. Freud merely chooses to substitute one text of experience for another in a traditional allegory, which valorizes the second text and disregards the first. In this we can see a conception of cognitive activity like that of Freud's contemporary, Alfred North Whitehead, who describes cognition as the discovery of "a pattern with a key to it" (1967: 26). Here Freud's reading of the dream is an instance of resistance to the very semiotic activity of dream interpretation that, paradoxically, he inaugurated and intended to promote in *The Interpretation of Dreams*.

Lacan's aim in Book II of *The Seminar* is to interpret the dream of Irma's injection beyond the impasse of Freud's resistance, which entails semiotizing the dream as an act of cognition, treating it as a narrative text more consistently than Freud was willing to do. But it also entails, we are suggesting, politicizing the text. In his rereading in *The Ego in Freud's Theory and in the Technique of Psychoanalysis*, Lacan reads the same dream segments as Freud did—the examination of Irma's throat, Otto's sexual transgression, Irma's contamination, etc.—but focuses on Freud's resistance to textuality as a semiotic element of the dream, as an *activity* of cognition. He seizes particularly on the challenge in the first half of the dream to Freud's therapeutic expertise and control when Freud looked into Irma's throat and saw her morcellated flesh. What Freud saw there as a hieroglyphic—"a big white patch . . . whitish grey scabs upon some remarkable curly structures"—is what Lacan calls a "horrible discovery," the great haunting of psychoanalysis (and cognition more generally) by ghostly tissue, the "talking cure" plagued by the body, not because the body escapes signification, the potential of mastery by cognition, but precisely because this text of conflicted meanings—"the flesh one never sees," the flesh "as it is . . . formless" as Lacan says (1991: 154)—demands representation and recognition *within* the structures of language and cognition.

The hieroglyphic of Irma's throat in Freud's dream, in other words, does not lie outside the cognitive province of language at all but precisely exemplifies language—the inscription of an absence and a call to enunciation that can be answered only with the further reiteration of cognitive linguistic structures, more language. Freud's response to the text of Irma's injection, however, is a gesture that foreshadows much of twentieth-century thought. He simultaneously "reads" and "misreads" the text in what will be the model for all performative and noncognitive dream interpretation, what Freud calls "working through." The misreading is most interesting because Freud seems to erase the subject of this dream enunciation—Freud himself—from the interpretation in an effort to make the special case of his dream the general case of an interpretative model.

For his part, Lacan chooses to focus on Freud's evasion of the throat as something to be interpreted, as an object of cognition. He reads Freud's assumption that the hieroglyphic of Irma's throat is a single and unified key to this dream, the "origin" of it and, consequently, the "model" of dream interpretation in general, as itself subject to the more general interpretative framework in which cognition is situated. In focusing on Freud's evasion, Lacan sees that the throat segment textually marks where the dream penetrates what *cannot* be known, what is not susceptible to the structured generalizing reductions of cognition, and this is precisely the point of Freud's strongest resistance to the dream, the part he makes biographical and the point where he closes down interpretation—his own denial of textuality and cognitive activity. To do this, Lacan self-consciously narrativizes his analysis; he ventriloquizes Irma's throat with the voice of psychoanalysis itself. Lacan's narrative presents an allegory of cognitive meaning very different from Freud's precisely because it narrates the scene of cognition. In that narrative, Freud looks down Irma's gaping gullet, and it says to him, "*You are this, which is so far from you, that which is the ultimate formlessness*" (1991: 155). Lacan interprets the dream, in short, as conveying the fact that meaning gets constituted only when it breaks down, when it marks *and* fails to mark an Otherness within itself.

Cognition and Oppositional Discourse

We have not traced Lacan's response to Irma's injection to give yet another explanation of the "Other," or of Lacanian metapsychology, or of how signs work. We are arguing, rather, that we go "back to Lacan" to find something we do not always see when we think of understanding and cognition: namely, the ideological critique—like something stuck in the throat—inherent in his "oppositional discourse." For example, Lacan's rereading of Freud's dream, semiotizing and resituating it *as a text,* as a *constituted* object of cognition, also demonstrates the synthetic nature of Lacan's approach to narrative cognition. By "synthetic," we do not mean artificial, of course, but rather an approach or understanding of narrative that connects with and theoretically allows for the acknowledgment of an "Other" reality beyond itself and its "knowledge." Lacan's recognition of failed meaning and the incommensurability of the "Other" are the openings to a world that is neither linguistic nor capable of cognitive apprehension—but which is also not a narcissistic dream. Lacan's "synthetic" approach, his respect for the "Other," contrasts with Freud's "analytic" and confident reading of the Irma dream; it opposes cognitive formalism and the mastering of an "original" experience the dream supposedly is about outside the structures of language and cognition. The whole point of the Lacanian reading, we are saying, is precisely to break the circle of the cognitive formalism of psychoanalysis, above all to escape the seduction of experience as entirely controlled by the ego. Lacan's point in relation to psychoanalysis—like our earlier point in relation to Darwinian "individualism"—is to show performatively, as Robert Young says, "that the ego is not a totality that can assimilate unconscious processes" (1981: 177). Instead of ego, Lacan speaks of the subject as an economy of relationships, which in turn presupposes *cultural* value bound up with (egoistic) acts of cognition.

In pursuing this analysis, Lacan breaks the hold of ego-centered formalism (and object-centered cognition) by treating the ego as no more than a signifier interacting with other signifiers under the control, as Freud says, of certain syntactic laws that constitute the unconscious. These signifiers, further, stand before an "Otherness," a failure of signification and cognition, that is not representable or controllable by language. This Otherness is especially evident in Lacan's analysis of the Irma dream when he identifies four agencies, or positions, he calls the Schema L. The first position is the role taken up by those who speak in the dream and whose voices focus our attention at any one moment—Freud himself, Dr. M, Leopold, and so on. Second is the role of an "other," or unknown, object in the dream—the unknown object of cognition—such as the enigmatic throat that Freud tries but fails to decipher. Third is the absolute sense of Otherness in the dream, what Paul de Man called "error" and Derrida calls "différance," the unrepresentable force of difference that structures (in the form of binary oppositions) and scrambles (in the arbitrary articulation of those oppositions) the dream as a narrative. And fourth is the resituated sense of what the dream means once the other positions have effected a new "reading."

This set of relationships (Lacan's Schema L) takes the form of Greimas's semiotic square, which can be depicted as follows in relation to the Irma dream:

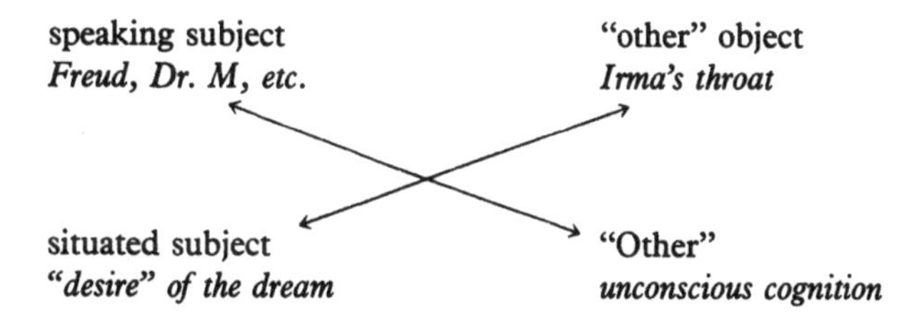

Lacan's inscription of the Schema L into the positions of his reading of the Irma text points up an intellectual tradition we examine in the final chapter within Western descriptions of cognition that thematizes the binarity we have encountered in this book by describing it as ideological opposition—the articulation of an "oppositional square." This theme can be traced from the pre-Socratics and Aristotle through Hegel, Marx, Antonio Gramsci, Jean Baudrillard, Robert Blanché, A. J. Greimas, Edward Said, and Fredric Jameson. This tradition of oppositionalism formulates specific strategies for thinking about what it means for ideas to oppose each other and at what point opposition becomes neutralized and then appropriated by the opposition. This tradition, in short, focuses on the workings of conflict within the "effects" of cognition. It articulates how to understand seemingly transhistorical cognitive truths in accordance with the application of the "syntactic laws" of cognition in specific historical settings. Our placing of Greimas in this setting, like Lacan's situating Freud in this tradition, makes his semiotics cultural. For this reason, Lacan has argued that "psycho-analysis is neither a *Weltanschauung* nor a philosophy that claims to provide the key to the universe. It is governed by a particular aim, which is historically defined by the elaboration of the notion of the subject. It poses the notion in a new way, by leading the subject back to his signifying dependence" (1978: 77).

The oppositional critics we are connecting with Lacan, moreover, tend to assume, as Lacan does, a common foundation for language and politics. Like Aristotle before them, they assume the dynamic nature of transformations that govern the possibilities for opposition in any cultural practice. In Aristotle's practice (elaborated in *On Interpretation*)—which is the prototype of Greimas's semiotic square—the attempt to understand opposition questions why conflicting interpretive possibilities move along particular lines and not others. If such *particular* conflict did not arise, oppositional critics traditionally assert, change—and especially important political changes—could not be understood, "accounted for," subject to cognition (Anton 1957: 86). Without the understanding generated by the oppositional metaphor, differences would be lost in the ongoing iteration of heterogeneity and chance and would have no ideological markings—no signs. "There can be no knowledge," as Aristotle warns in *Posterior Analytics*, "of that which [exists or comes to be] by chance" (1981: 42). Aristotle sees the concept of opposition, as Baudrillard and Greimas affirm after him, as embodying the very possibility of rational analysis and of what could be termed political "cognition" in reading experience.

Lacan's Schema L and his approach to the Irma dream reflect these analytical assumptions, as do, less self-consciously, the various Greimassian squares we have presented throughout this book. In the classical oppositional square, as articulated by Aristotle in *On Interpretation* and by Greimas in "The Interaction of Semiotic Constraints," there is a first-level "contrary" relationship, as there is in Lacan's opposition between the speaking subject (the "I") and the "object other." These are in opposition not as a (contradictory) negation of each other (as in the relationship of *A* and Not-*A*) but as regards something *other* than ("contrary" to) the speaking subject—non-*A*. Jakobson and Greimas describe this relationship as that between the extreme poles on a single axis—"male" and "female," for instance, on the axis of sexuality; or "subject" and "object" on the axis of discourse.

On a second level on our diagram of Lacan's Schema L is the radical

"Other"—the absolute Otherness of difference and dissemination—in a contrary relationship with the resituated subject. In this two-tier formulation of the Schema L, Lacan reads the subject/other opposition exclusively as an articulation in a set of cognitive relationships, and he demonstrates that the economies of relationship form hierarchies in an ideological framework. This happens as each position of the Schema L further defines a hierarchy of values and maps the pathways of power in cultural cognition. This process is particularly visible when a culture chooses arbitrary representations for the "Other" such as "Father," or what Lacan calls the "Name-of-the-Father." Such misrecognized coordinates (or "axes") of understanding, as Lacan demonstrates, are inevitable only in that we cannot interpret without them. They must be there if there is to be cognition—which is now understood as an always positioned or particular *ideological* understanding. They have the effect of rendering cultural discourse as readable by constituting experience within narrative articulations which themselves situate cognitive activity. The result for Lacan is an economy of values inscribed by the Schema L's positions, and the particular manner of signifying positions—articulated according to the "syntactic laws" made evident in the Schema L's operation—formulates an instance of ideology. That is, the complicated positions suggest a critique of the value that seems to be inherent, self-evident, within the cognitive understanding of a particular narrative at a particular moment.

The fourth term completes the ideological critique implicit in the Schema L with what Lacan calls the *moi*, the historically situated subject. (The whole Schema L represents Lacan's sense of the subject.) This position suggests a new interpretation, a transformation of the axis upon which the "logic" of cognitive apprehension takes place. At this farthest reach of the square, the new "subject" both completes the ideological values the square represents, and it marks the *instituted and constructed* nature of that seemingly self-evident understanding. This fourth position, in effect, goes beyond the other oppositional relations by formalizing the reconception of their relationship. As we will argue in a moment, it demonstrates that what seems to be natural and inevitable is, in fact, always a *special case* of the activity of understanding. Whereas the first-level relationship signals the coexistence of "contrary" possibilities, and the third position articulates the values inherent in the square, the fourth has no clearly oppositional relationship with the first. The traditional name of the second level in relation to the first is "subaltern." In Greimas's semiotic square of actants we presented in Chapter 2, this fourth position is inhabited by the helper actant which embodies the power rather than the knowledge of the hero.

In this instance, the fourth position inscribes power within cognitive relationships. It creates the possibility of reconceiving the nature of all the relationships the square describes so that the seemingly natural subject conceived as a subject of knowledge—a cognitive subject—is seen to be a special case of the wider category "subject of knowledge/power." (This is precisely J. L. Austin's logic in *How to Do Things with Words* when he concludes that constative discourse is always a special case of performative discourse [1962: 139; see also Culler 1982: 110–34 for a discussion of the Derrida-Searle debate on this issue].) In the articulation of "agedness" in the semiotic square of Chapter 3, this position is inhabited by the antipositivist position of "negative growth," the *situated* value of aging. In both these cases, the fourth position foregrounds a potentiality made possible by the range of reference created on the first level.

In Lacan, the fourth position also marks a possibility newly generated by the relationships of the Schema L. His reading of the subject/other opposition demonstrates the economies of relationship engendered by binarity, especially the economy of ideological understanding. This happens as each position of the Schema L further defines a hierarchy of values and maps the boundaries of a cultural text. These coordinates of an ideological reading may seem mechanically to program merely formal and artificial possibilities. Such oppositional "systems," however, as Anton says, are inevitable if there is to be understanding at all, which is now seen to be an always ideological understanding. That is, they have the effect of rendering cultural discourses readable by submitting "all the differences to . . . the principle of contrariety, which in turn becomes the pivot-point for relating, organizing, and systematizing differences" (1957: 86). The result for Lacan is an economy of values inscribed by the Schema L's positions—themselves "escaping" the unintelligibility of "mere" differences, the ultimate unintelligibility of relativism we will examine in a moment—within a nonpoliticized field of heterogeneity. Finally, reading the pattern of these positions according to the "syntactic laws" made evident in the Schema L's operation articulates the place of ideological understanding—that is, "cognition" itself—within narrative.

What we get with Lacan's analysis of this dream, as is true with Foucault, Jameson, Said, and the oppositional critics we examine in the final chapter, is the activity of semiotic and ideological analysis in cognition itself. Ideology, in this view, is not an immutable system of belief but an "effect" of discourse produced within a system of differences that are continually subject to transformation within that system and, ultimately, within history. One cannot arrest the cycle of potential transformations of a historically bound—and, thus, dynamic—subject. The "split" in the subject, so important to Lacan, is a split or a doubleness in relation to language and cognition. It is, as Greimas suggests, a "split" between the subject as *subject over* and *subject to* knowledge, between the activity of cognition and the grammar of the sentence in which it appears. In other words—in Greimas's actantial terms—the split is that between sender and subject, what he and Courtés later call the opposition between the "enunciator" of discourse and the actor within discourse. For this reason, as we saw in Chapter 2, the receiver is so important to Greimas's description of narrative cognition. And, as we saw in Chapter 3, the sender is so important to an empirical description of narrative cognition. In these terms, the subject is the subject of consciousness, and the sender is always, to some degree or other, beyond consciousness. Such "unconsciousness" is not personal: it is precisely *cultural*, just as the sender, in Greimas, is always larger than the discourse it is associated with even though, as an actantial role, it does not actually transcend the narrative in which it is embodied. Thus the situating of the subject in the Schema L—and in the articulation of cognition in Lacanian psychoanalysis more generally—is a cultural event, the institution of cognition that articulates and disrupts the ideological frame that produced it. As a cultural phenomenon, the situated subject marks understanding historically as the diachronic emergence of seemingly transcendental cognitive "truths."

We are not claiming that Lacan follows Aristotelian and Greimassian protocols to work out oppositions in a Freudian metapsychology. Lacan has understood, however, as Aristotle and Greimas in their own ways also posit, that the relations of opposition map a logic of resistance within the ideological constitution of cognition and understanding. In his discourse on the Irma dream, Lacan implicitly formulates the ideological construction of narrative meanings—the manner in which cognition presupposes *value*. He shows that the power of difference is manifested in the power to narrate and that this power—strictly speaking—belongs to no one nameable. Accordingly, the empowerment of the father, like the arrogance of Freud's surrogate Otto in the Irma dream—and like the embodiment of the sender in any narrative—is precisely an ideological maneuver to be enacted within and not to be accepted as in some way "beyond" the narrative in which it appears. The "natural" empowerment of the father, or any agency, must be taken as a political use of language and cognition—a suturing of authority to a position it is not "naturally" connected to, a suturing of the "phallus" as a signifying authority to the father's body. The "phallus," for Lacan, is the constituted object of cognition which, "sutured" in this way, is apprehended as simply and self-evidently "natural." Such "natural" authority is also seen both in the deference, in some cultures, paid to the elder as sender and in the disregard, in others, of elders as people whose power of narrative articulation has been replaced by other institutions.

It is exactly the empowerment of the father in psychoanalysis, the male will to (psychoanalytic) power over women, that Lacan calls into question by providing the instrument—in the oppostionalism of the Schema L—to question it. In this sense, Lacan is positioned *against* Freud and *for* psychoanalysis and the psychoanalytic understanding of language and culture. In this sense, the Lacanian narratology we describe in the next chapter, derived as it is out of the Schema L, is itself inherently a political construct, a critique of ideological and cultural investments and a shifting of authority (despite the claim of some theorists to the contrary) away from "natural" and privileged authority—the authority of "objectivity."

The Social Construction of Opposition: Relativity vs. Relativism

Before we turn to the Lacanian understanding of narrative, we will examine another narrative critique. Bruno Latour offers a description of the social constitution of cognition that takes as its focus Freud's contemporary, Albert Einstein, and the presentation of his discussion can clarify what we are calling the critique of cognitive objectivity. In "A Relativistic Account of Einstein's Relativity," Latour subjects Einstein's 1920 "semi-popular" book, *Relativity, the Special and General Theory,* to a sociological analysis. He examines the book in relation to what Greimas and Courtés call "shifting out" and "shifting in" of the enunciator of language. Shifting out is the process in language that shifts attention away from the enunciatory situation of discourse—away from the sender-receiver relationship in which someone is speaking to or "communicating" with someone else—and toward the configuration of actantial relationships within a text, the organized "content" of the discourse-communication. It is, as we have already noted, the shifting of understanding from the series of distinct messages—what Greimas calls the presentation of "a series of messages as algorithms"—to an actantial "simulator of a world from which the sender and the receiver of a communication are excluded" (1983b: 134). Such a shifting draws attention to the level of "knowledge" rather than "power." It presents as self-evident the authority of "objectivity." Shifting in is the opposite process of calling attention to the situation of enunciation, as when a narrator reminds us that *someone* is speaking by using a first-person pronoun.

In Latour's analysis of Einstein's narration of relativity theory, he uses the binary opposition between shifting in and shifting out to examine the work of Einstein's text and the social institution of scientific knowledge more generally. In doing so he is crossing the "levels" of an oppositional square (although he doesn't articulate his argument in terms of a semiotic square).

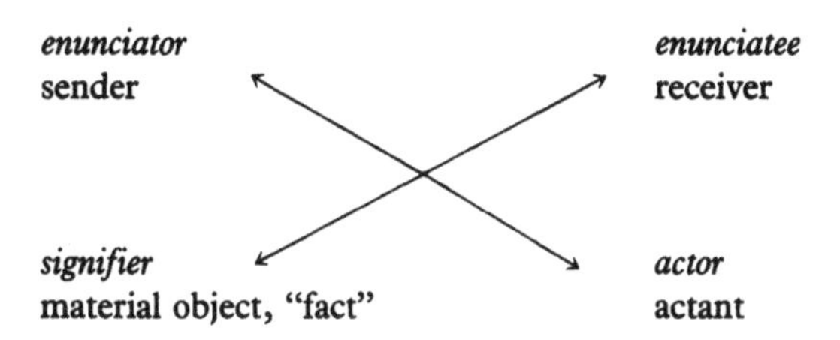

This square articulates the contradictory opposition between the level of enunciation and the level of signification. By describing the opposition between enunciator and actor in Einstein particularly and in science in general, Latour is describing science as a special case of the situated cognitive activities of shifting in and out in cognition, just as Einstein himself describes Newtonian physics as a special case of physics, and Lacan describes Freudian clinical psychoanalysis as a special case of the psychoanalytics of cultural ideology.

In specific terms, Latour demonstrates that the shifting in and out of the enunciator/sender of discourse is precisely the *focus* of Einstein's account of relativity, and it is precisely the general cognitive case of which Einsteinian relativity is a special case. In this work, Latour argues, Einstein "is not only obsessively interested in the staging of the very frames of references that allow spatial and temporal shifting-out, but he also focuses on the shifting-in. As I have said, it is not the former but only the latter that creates distinctions between fiction-writing and fact-writing" (1988: 12). Shifting in is a return to the enunciator, the "sender" of a message whose *position* allows the possibility that infinite semiosis can be contained within a frame of reference. Thus, Einstein's focus on shifting in creates the difference between relativity and relativism. Latour notes that "relativity . . . is the exact opposite of relativism" (1988: 14). In a world of relativity, the "delegated actor" in the narrative discourse from which the sender and receiver are excluded "has no personal point of view" so that the number of delegated subjects of cognition, no matter how large that number is, will have "no privileged point of view; which means that no matter how far away I delegate the observers, they all send back superimposable reports that establish my credibility; which means that it is possible to escape from fiction" (1988: 14).

Of crucial importance here is the possibility of "superimposable" reports, reports that can be aligned so that fact-writing—Latour's articulation of the activity of cognition we have pursued in this book—can, in fact, take place. "What appears confusing in Einstein's text, as well as in the opposition between relativism and relativity," Latour notes,

> is this apparent paradox: if there exist many points of view each claiming to be privileged, no one of them can get an edge over all the others; if, on the contrary, there are no privileged points of view, this means that there is nothing to prevent one of them getting an edge over all the others. . . . It is the same paradox as that of [economic] liberalism. As long as any movement of goods, money or people is interrupted by many local franchises, protections, tariffs, feudal systems, particular regulations, traditions, irreducible cultures, it is impossible to capitalize on any large scale. (1988: 15)

The relativism of points of view—like the relationalism of the corners of Greimas's semiotic square—creates the possibility of the destruction of relativism and the institution of relativity. Here is the chart Latour offers to describe the opposition between relativism and relativity (1988:16).

Relativism *versus* Relativity

Relativism	**Relativity**
Privileged points of view	No privileged points of view
Independent observers	Dependent Observers
Unequivalence of observations	Equivalence of observations
No superimposition of traces	Superimposition of traces
Enunciator has no privilege	Enunciator gains in the end
No large-scale privilege	Large-scale privileges
No possible omniscience	Omniscience is possible

This opposition makes the "effect" of cognition possible by creating, as Einstein does in relativity theory, the possibility of an "effect" of omniscience. Einstein does this by making the *activity* of the construction of frameworks of understanding—frameworks of shifted-out articulations of observation—the focus of his work, just as we have been emphasizing, throughout this book, the *activity* of cognition. For Einstein, as Latour describes it, "there is no longer any one frame that might be used as a rigid and stable reference, into which confidence is vested; confidence is now put into the transversal link that allows all frames, no matter how unstable and pliable, to be *aligned*" (1988: 17). That is, the effects of omniscience governing cognition are socially and culturally determined.

For this reason, Latour argues, "what semioticians call without further ado 'shifting out and in', because they mostly consider narrations that are content to be read as text and fiction, is offered a precise meaning by Einstein because he studies narrations that he wants to *distinguish* from texts and fictions" (1988: 18). That precise meaning is the *social activity* of science, of cognition, of articulating the sender of discourse that allows and accomplishes the alignment of instituted understandings. "In this book," Latour says,

> Einstein's fiddling with time and space does not lead, as we can now see, to the metaphysics often triggered by his writings, but to an *infra-physics* of crucial importance for the sociology of science. Instead of frames of reference, we are presented with the practical work of setting up frames; instead of characters, we now see the hard work of disciplining and managing delegated observers and instruments; instead of taking information for granted, the encoding and decoding of information are now made visible. Inscriptions, subscriptions, transcriptions: the word 'relativity' refers to this lowly work of building and relating frames to one another in such a way that some kind of stable form can be maintained which can, then, be cumulated, combined and superimposed at some point. (1988: 20)

In this way relativity allows for the social institution of knowledge. The difference between fiction-writing and fact-writing, Latour says, is the possibility and the need to account for the referent. In Lacan's account of the Irma dream, as we have seen, Freud prematurely turns to referents to set up a frame of reference for interpreting the dream, and he does so without narrating the practical work of setting up such frames. Lacan, on the other hand, and Continental semiotics more generally, too quickly dispense with such referential frameworks, taking the special case of fiction-writing to be the defining case of cognition.

Latour argues that we should take the social constitution of science as the general case of semiosis, especially in the case of Einsteinian relativity. "The two principles, that of Einstein's relativity and that of semiotics," he writes,

> are one and the same. They both state that to talk of an *external* referent independently of the structure of the report is devoid of meaning. They both state that we are always in between at least two frames and that the deeper we go into physics and cosmology the more we should examine the conditions of the narration that stage these frames. They both state that an effect of reality is built in by the superimposition of reports sent from at least two frames of reference to a third one.
>
> Why is the first one accepted with gratitude while the other is greeted with horror, by natural as well as by social scientists? It is simply that the opposition between relativism and relativity which is so clear in the case of Einstein has not been made as clear in the case of semiotics. The reason for this lack of clarity is to be sought through the question of the referent. (1988: 28)

Semiotics dispenses with the referent since reference itself, given the radically *relational* nature of semiotics and cognition, can *always* be shown to be an arbitrary stopping place in a process of infinite semiosis, an arbitrary act of substantification. In Greimassian terms, every semiotic square can be named and positioned within another square; every corner of a square can be shown to generate its "own" square. In terms of the examination of metaphors in Chapter 1, any "literal" articulation of "nonsense" can always be shown to be metaphorical within a relative frame of reference. In the same way, any "method" of investigation can be shown to be inextricably caught up in local "epistemological" assumptions.

Yet the fact that these things can always be done—that the "logic" of discourse and cognition can always be shown to lead to an aporia, and that seemingly objective cognition can always be shown to be situated historical events—does not have to lead to the abolition of reference. Rather, reference can be reoriented and reconceived. In Latour's analysis of Einstein's text, using the semiotic tools for analyzing texts without recourse to reference, he notes that "it is impossible to go beyond narration and beyond some superimposition of documents" in order to measure the cognitive content of the text—whether it is true or false (1988: 29). In other words, "shifting in" to find the sender—the authority of cognition—is always repeatable, always governed by relativism. Yet in fact-writing rather than fiction-writing such shifting in can create what he calls an "underwritten referent" (1988: 29). That is, the semiotic activity of shifting in can self-consciously *institute* a sender and thus institute cognition by means of the social activity of semiosis.

How can we distinguish L. S. Feuer's book *Einstein and the Generations of Science*, Latour asks, from a work of fiction? "Can you shift in all the way back to Feuer's office," he asks, "and superimpose in some way the documents he mobilizes in his text with others? If no, then the boundary of the narration is such that you have only a text; from the text to the enunciator there is a *non sequitur,* a gap" (1988: 29–30). In this case, as Lacan argues, the subject of knowledge is always articulating, unconsciously, what he has called "the discourse of the Other." This discourse is the discourse of the sender of whom the subject of knowledge is *unconscious*. In the same way, in Greimas, the individual subject always unconsciously repeats what has already been articulated by a sender who offers the very paradigms—the cultural paradigms—by which cognition takes place. And more: the sender itself is relative; it exists within the relativism of the frames of reference it authorizes so that it can be seen to be, indiscriminately, the sender of grammar, of semiotic squares, of binarities of cognition, and so on. The very existence of an authorized sender is simply the textual "gaps" that Latour mentions and that we will see in Lacan's implicit theory of narrative. From the point of view of the *singular* subject of cognition, just as from the point of view of the singular author of a text, text production—cognition itself—is fiction-writing.

But Latour suggests there is another way to answer whether or not one can shift in all the way back to Feuer's office and superimpose in some way the documents he mobilizes in his text with others. If this is possible, Latour continues, then

> the boundary of the text is stretched further in; there is continuity, a network is in place. But who does such a verification? Who goes to the office of the writer to check this ultimate superimposition? *Another scientist,* another writer who is busy expanding still another network by establishing a continuous link between the inscriptions mobilized in his text and what a potential reader could wish to see in his office, were he to check, and so on. In other words, there are three things we cannot escape from: discourses, inscription devices and networks—that is, infra-physics. (1988: 30)

In this understanding, the semiotic analysis of discourse is just a special case of the scientific cognition of underwritten referents. This is the special case of "fiction" in which Latour's elements of relativism—the privileged point of view of its author, the lack of privilege of its narrator (enunciator or sender), observations that are independent of other observations, an enunciation that is seemingly unique (without the superimposition of traces of other enunciations), the impossibility of omniscience beyond the confines of its own discourse—are the marked special cases of the general case.

The general case is that of "relativity" in which textuality is social and continuous—Latour's "network"—without the plague of an unbreachable "gap" between discourse and sender. Here, the referent is underwritten—cognition itself is underwritten—precisely by the sender of a discourse in which no author is privileged, but each observation is dependent on other observations. The form that dependence takes is the possibility of superimposition of traces of one enunciation on another until a privileged narrator (such as the *complicated* sender that we described in the previous chapter inhabiting the role of elder) is instituted that authorizes the omniscience of cognition beyond the special case. In this narrative version of Latour's argument concerning the socially underwritten cognitions of science, we can see that fiction indeed is a special case, just as, in Latour's argument, Einstein's text is a special case of the activity of understanding.

It should also be clear now that the crucial activity of superimposing traces of enunciatory activity is precisely what our examples of Greimas's semiotic square and the focus on aging accomplish in the superimposition of "levels" of binarity and "stages" of development. This activity of superimposition can be seen in Freudian "overdetermination." Such superimposition, as we have followed it in Lacan's Schema L, articulates culture and cognition. Speaking of Hegel, Jacques Derrida describes the function of negativity in ways that are superimposable on Latour's description of the sociology of science and his earlier description of the status of contradiction as "neither a property of mind, nor of the scientific method, but . . . a property of reading letters and signs inside new settings" (1986: 20). "In discourse," Derrida writes, "(the unity of process and system) negativity is always the underside and accomplice of positivity. Negativity cannot be spoken of, nor has it ever been except in this fabric of meaning. Now, the sovereign operation, the *point of nonreserve*, is neither positive nor negative. It cannot be inscribed in discourse, except by crossing out predicates or by practicing a contradictory superimpression that then exceeds the logic of philosophy" (1978: 259). Such a contradictory superimpression—like Latour's superimposition of traces and Lacan's superimposition of conscious and unconscious subjectivity in cognition—describes and narrates the functioning of semiotic, oppositional squares and the understandings instituted by their logic and activity.

Moreover, it describes *the activity of referential knowledge* that empirical science presents as already given. In her analysis of Lacanian psychoanalysis in terms of the performative aspect of language, Shoshana Felman describes such cultural cognition as a "*change in status* of the referent as such" (1983: 75). "Contrary to the traditional conception of the referent," she writes,

> referential knowledge of language is not envisaged here as constative, cognitive knowledge: neither for psychoanalysis nor performative analysis is language a *statement* of the real, a simple reflection of the referent or its mimetic representation. Quite to the contrary, the referent is itself produced by language as its own *effect*. . . . This means that between language and referent there is no longer a simple opposition (nor is there identity, on the other hand): language makes itself part of what it refers to (without, however, being all that it refers to). Referential knowledge of language is not knowledge *about* reality (about a separate and distinct entity), but knowledge that *has to do with reality*, that acts within reality, since it is itself—at least in part—what this reality is made of. The referent is no longer simply a preexisting *substance*, but an *act*, that is, a dynamic movement of modification of reality. (1983: 76–77)

In Lacan, that is, as in Latour, the sender can be articulated as a category of culture—a category of cultural cognition—so that the discourse of the Other has to do with reality, just as Einsteinian delegated observers are engaged in doings with reality. In fact, Lacan's theory of narration, implicit in his description of the unconscious as the discourse of the Other, allows "Truth" itself to find voice. It is to that theory that we now turn. Then, in Chapter 6, we return to the special case of the discourse of old people and "shift out" to the rhetoric of that discourse. This emphasis enables us to situate the knowledge of old people in relation to the cultural-discursive subject of that knowledge in a scientific inquiry whose empiricism is reconceived in the light of the activity of cognition.

5

PSYCHOANALYSIS AND NARRATION

Lacan, Greimas, and the Desire of Cognition

Lacan and Narration

Jacques Lacan's concern with the Freudian subject, like Jerome Bruner's concern with the subject of cognition, suggests a position in regard to narration—an approach to, and an articulation of, a relationship between cognition and narration. It is an approach derivable from Lacan's view—his central insight into psychoanalysis—that *the unconscious is structured like a language*. It says simply that narration, too, operates like a language, is a language, and manifests cognitive-linguistic operations in various ways. Narration exists, finally, within the context of an unconscious "discourse," within the bounds of what Lacan calls the "discourse of the Other." Since the early 1970s, though earlier in France, literary theorists have sought to understand this central insight of Jacques Lacan's rethinking of psychoanalysis. In a sense, they have attempted to reverse it, not to disprove it necessarily, but to grasp how language in narrative texts is constituted, buoyed up, permeated, and decentered by the unconscious. The aim has been to understand (reversing Lacan's statement) how "*literature*," in Shoshana Felman's words, " . . . *is the unconscious of psychoanalysis*" (1977:10). In this regard, this chapter, with its focus on the understanding of narration in psychoanalytical terms, attempts another step in situating narration as an effect or product of *cognitive* activity, the cultural activity of understanding. In this frame, narration is in a sense already psychoanalytic, even before or apart from, even despite, both Freud and Lacan's semiotic rereading of Freud.

The theoretical melding of the unconscious and language—psychoanalysis and linguistics—does take place *before* Lacan, as Lacan himself insists, in Freud's *cognitive* enterprise, particularly in *The Interpretation of Dreams* (1900), *The Psychopathology of Everyday Life* (1901), *Jokes and Their Relation to the Unconscious* (1905), and many of the metapsychology pieces. Freud, unlike Lacan, did not have access to Ferdinand de Saussure's formulation of semiotics on which modern linguistics is based (the *Course in General Linguistics* was not published until 1916, sixteen years after *The Interpretation of Dreams*). But, as we saw Lacan demonstrate, Freud was simultaneously "discovering" the sign and scientific semiotics for psychology just as Saussure was doing so for linguistics, and Russell, Wittgenstein, Turing, and others were doing so for Anglo-American cognitive science. Because Freud's major work on dreams "appeared long before the formalizations of linguistics," Lacan argues, psychoanalysis even "paved the way [for linguistics] by the sheer weight of its truth" (1977b: 162).

This discovery of Freud's cognitive semiotics can be seen, as it is in Irma's dream, in a another passage in *The Interpretation of Dreams*, a central Freudian "scene of writing" that Lacan discusses, in which Freud describes two different modes of the cognitive apprehension of narrative. In this passage Freud gives a sample dream about a "house with a boat on its roof, a single letter of the alphabet, [and] the figure of a running man whose head has been conjured away" (1965: 311–12). Now as Freud interprets it, or as Lacan might, this brief narration is actually about semiotic cognition. We see this in the placement of a house with too much of a top (a head) in relation to its contrary, to the image of a running man "whose head has been conjured away." The images form a binary opposition. There is an excess of presence in the boat-head, and there is a stark absence in the man's missing head. In this contrast, further, is a theatrically distinct opposition between presence and absence, the condition of difference in a binary system that makes a sign (and cognition) possible.

That is, what stands apart from and yet represents a semiotic difference is the possibility of representation itself, the signifying possibility (as represented in the dream) of "a single letter of the alphabet." This signifying capability, moreover, exacts a human cost in that the subject of language is signified within and, simultaneously, is alienated from language—hence the man's symbolic castration in being headless. Such alienation within cognitive activity is the mark of Freud's contribution to the study of cognition beyond the logic of semiotics and the positive empiricism of Anglo-American cognitive science. In this case, Freud speaks, as does Lacan, in several registers simultaneously. In both his choice of an example and his commentary in this dream analysis, for example, his discourse foregrounds the structurality of language and understanding. The dream's figures or elements—Lacan calls them "sliding signifiers"—move over certain "positions" in the dream in a kind of articulation structured *in* language and, through interpretation, actually *about* language. That is, they inhabit the "binarity" of language, semiotics, and cognition itself.

Freud's main interest in this "scene of writing" in *The Interpretation of Dreams* is interpretation itself, the principles by which we read the dream as essentially a picture, at one extreme, or as a narrative, at another. His first approach to the dream focuses on the "*manifest* content" (1965: 311) of the narrative details as if they are in a "pictographic script" (1965: 312) in which, as in the models of cognition we have examined, all of the pieces are present simultaneously for inspection as in a "picture-puzzle, a rebus" (1965: 312). In this pictographic approach the disjunctive quality of the picture—"house with a boat on its roof," etc.—creates a problem for interpretation. The dream elements, dislocated from a familiar context and yet fixed in place, have no significance except possibly as icons with preexistent and fixed meanings. If they are icons, the dream narrative may be read as an intelligible picture-story, a deciphered rebus with its meaning transparently available. For the second mode, instead of being a picture, the dream functions like a language governed by, as Freud says, "syntactic laws" (1965: 312). In this approach, by contrast, the dream elements are not seen, at least not exclusively, as if in a picture or as inherently meaningful. Rather, they are now arbitrary "signifiers" occupying "positions" in discourse defined in terms of binarities with no intrinsic or assigned meaning. In this "linguistic" approach, Freud explains, we "replace each separate element by a syllable or word that can be represented by that element in some way or other" (1965: 312). This kind of interpretation, in other words, takes place through the substitution of one element for another—through a precise allegorical activity—according to certain culturally determined narrative codes. Crucial here is the interchangeability, the substitutability, of dream elements—signifiers—which can hold particular places in the dream. The dream is an interpretable text only according to the possible substitution of elements. While in Freud's pictographic approach particular dream elements are meaningful (or lacking in meaning) as they are locked in a particular pattern, in the linguistic approach dream elements lack inherent meaning but can stand in for (be replaced by) other elements within a certain structure or set of possibilities.

In view of the choices—the interpretive dilemma—that Freud sets up in this discussion, we can see that Lacan chooses to go the second way of interpretation, that of language and linguistics, the path, he says, of "writing rather than of mime" (1977b: 161), in the same way that cognitive science in general has pursued the methodology of binarity rather than the epistemology of fixed transcendental truths. The apparent simplicity of such a choice, however, is deceptive. Freud's pictographic approach, on closer examination, actually contains important aspects of the linguistic approach and so cannot be completely discarded. If the dream elements are not icons, Freud's picture-dream is reduced to being a structure of juxtaposed elements. Because the particular elements themselves are without meaning, what is left are the combinatory possibilities of dislocated and yet copresent elements—contiguous and/or simultaneously adjacent to each other. In the second, or linguistic, approach, Freud emphatically does not attribute inherent meaning to the dream elements. He does say that their relative positions, their existence in a system of differences, affect another process, that of elements being selected to go in the dream initially and to replace other elements as interpretation takes place, especially to replace the "original" (or manifest) elements that would otherwise form a pictograph. In short, the pictographic approach, as Freud describes it, emphasizes combinatory possibilities; the linguistic approach, as we are using the term here, indicates a selective process of one element substituting for another. In both approaches, elements in a series are placed in different relationships, but in being so placed they are also being selected and then substituted so as to represent any number of combinatory possibilities.

This is the distinction of post-Saussurean linguistics in which "combination" emphasizes purely relational concepts such as syntax, concepts inconceivable, however, apart from the selective (or paradigmatic) possibilities of words and word parts, such as morphemes, available in language as a whole. Roman Jakobson insists—to point up the interdependence of combination/selection as a single system of discourse—that in any one "utterance (message) is a *combination* [syntagmatic dimension] of constituent parts (sentences, words, phonemes) *selected* [paradigmatically] from the repository of all possible constituent parts (the code)" (Jakobson and Halle 1971: 75). Thus, while Freud for his own purposes separates these analyses into two approaches to understanding, insofar as they represent syntagmatic and paradigmatic analyses, they are interdependent and imply each other. Pictography and linguistics, Freud's commentary aside, are constitutive elements of cognition rather than different modes of analysis.

And yet, and perhaps this is Freud's real point, such extremes of analysis—especially for narration—do exist usefully as ideals. We surely can imagine a methodology of understanding that, merely tending toward the ideal (the "lure," as Lacan says) of naturally meaningful pictographic inscription, would emphasize the spatial forms of cognitive apprehension. This approach, of course, actually exists in the critical tradition associated with modernist fiction and spatial form. Joseph Frank imagines "modern" narrative structure as somehow escaping temporality, yielding its intended content when narrative elements are arranged as adjacent and copresent in the semblance of a picture, an organization Frank calls a nontemporal unity. In a similar way, the empirical experiments we examined in Chapter 3, attempting to isolate "the general nature of the phenomenon being described" (Myers 1990: 196), assume the nontemporal unity of "fact." Here narrative form resides inherently in the pattern of individual elements which are presented simultaneously as in a still life. At another extreme, we can imagine narration as tending toward being an operation in which elements, signifiers, stand in for other elements in a sequence. This possibility, too, is a reality. There is the tradition of modern narrative interpretation inspired by Saussure, Propp, and Marcel Mauss, and developed by Lévi-Strauss and Greimas, which continues on to Lacan and the poststructuralists: such a tradition sees narration as language process, not as a pictograph, but as a linguistic system of substitutions and alterations in time. Freud's discussion of the dream is, in fact, a rough but reliable map of the areas of cognition and narrative theory—pictographic and linguistic—that have existed and dominated definitions of understanding since 1900. It is Lacan's contribution to show that the operation of the unconscious, encompassing the extremes of what Freud calls pictographic and linguistic analyses, is itself a culturally determined linguistic process and to bring this insight intensely to bear in the psychoanalytic critique of the subject of knowledge and experience.

Lacan's interpretive choice, then, is not linguistics over pictography, or selection over combination—choices not available separately. It is rather a move to show the explicitly linguistic structure of the subject within an unconscious discourse. By "subject," Lacan means both the agency of knowing *and* the site at which this agency functions, has form, and is meaningful. Lacan's view here has implications for the narrative terms we have been using: what Freud calls the manifest content of a dream (or narration)—the pictograph—does not stand alone as a privileged structure. It is already a repetition, the result of a previous interpretation, of an already accomplished interpretive process—Freud's "dream work"—that already having taken place (and fated always to be a repetition) has produced the pictographic representation of a dream thought. Lacanian psychoanalysis takes this manifest content, this monument to a previous interpretation, and—treating it as one phase of interpretation—places it on narration's margin as a part and just one effect of the unconscious process involving metonymy and metaphor. As manifest content, this "literal" story or plot is "real," just as it is "real" when it displays the traces—the "gaps" in meaning or "lapses" of logic Latour mentions—that not only suggest the unconscious system that produced it but also indicate the social networking that covers (or "overwrites") those gaps and "underwrites" truth.

In this way, the so-called manifest content (really an "old" interpretation) is resubjected in interpretation to the same socially underwritten process of combination *and* selection, metonymy and metaphor, that produced the manifest content in the dream work initially. The cognition of narrative interpretation actually mimics and repeats the dream work analytically, as if to give the message back to its sender "in an inverted form" in narration (Lacan 1977b: 85). The actual dream work began with unconscious desires and dream thoughts and moved the thoughts through displacement (metonymy) and condensation (metaphor). Lacan's analysis of narration, as his readings of *Hamlet* and "The Purloined Letter" show, begins with language, the product of a previous cognitive activity, and proceeds to rediscover—by giving attention to certain "gaps" or "lapses," the indications of the unconscious process—the Other discourse that previously produced the so-called manifest text. In this way, narrative interpretation from a Lacanian viewpoint in fact does reverse Lacan's statement (as Lacan himself reverses it) about the unconscious and language and finds interpretation in the unconscious structure of language, which, in turn, is structured like a language, which, in turn—and so on. Freud's dream narrative, then, in this Lacanian sense, is *structured like a subject of cognition* in that it has the unconscious structure of language.

In this analysis, Lacan conceives of narrative as the work of the signifier as it can be known in its metaphoric and metonymic operations. Narrative conveys the fortunes of the signifier, its history, in relation to its own repressed origin in the unconscious cultural determinations of cognition. Narration—irremediably diachronic and synchronic—repeats and represents unconscious discourse in the only way the unconscious can be known: as a sequence of opportunities for linguistic substitution and (re)combination, an abstract model of cognition. The potential for continuity and unity in such sequences makes possible the "gaps" or "lapses" that indicate the "Other" scene of signification, the repressed scene of writing not a part of manifest narration but which, like other cultural institutions, holds it up and enables it to exist as a cognitive activity. In this formulation we are already assuming three fundamental propositions that characterize—though will not limit—a Lacanian concern with narration: Narration is structured to constitute a knowing subject in language. This places Lacan in a tradition of narrative theory with a linguistic orientation. Narration's manifest content is a product of the unconscious activity that is both the precondition of narration and the site of its appearance. This says essentially that the subject of narration, what gives it form and meaning, will always be other than what is signified *in* narration, or what is signifiable *as* narration. The subject of narration, in short, is the product of an interpretive activity. Last, the unconscious cognition of language and the processes of that unconscious cognition revealed in the "gaps" or "lapses" (inconsistencies, failures of speech and signification, etc.) that appear in a narrative's manifest text—the very moments when narrative as such seems to break down. This final proposition situates narrative interpretation, too, as a movement, a trajectory, a contingent effect, within the larger (unconscious) discourse of the Other. This larger discourse, the agency of repression, is the Other of cultural activity, a sender of discourse whose chief aim is *not* communication, but the articulation of "truth."

In practice, a Lacanian narratology based on these propositions is likely to be a triple reading: of the manifest text, of the unconscious discourse that conditions or buoys up the manifest text, and then of the reconstituted (repositioned) manifest text—a new interpretation. The *text* read here obviously is neither the New Critical, positivistic text—ambiguous and ironic but absolutely knowable—nor is it a rebus in which one reads and finds whatever one wants. To the question "Is there a text in this classroom?" we can answer yes. The Lacanian text is definitely *there* in the act of reading, but it is exclusively text-as-cognitive-production.

This narrative text differs from Greimas's structuralist one in that Greimas's communication model of sender/receiver (addresser/addressee) is absorbed by Lacan's formulation of a split subject and a set of "positions" existing *within* narration—not necessarily between sender and receiver. This shift—marking an important shift from structuralist to poststructuralist modes of thought—occurs largely because "the conception of language as communication," as Rosalind Coward and John Ellis hold, "tends to obscure the way in which language sets up the positions of 'I' and 'you' that are necessary for communication to take place at all" (1977: 79). The key phrasing here is "sets up the positions of 'I' and 'you'"—that is, "set up" positions through structuration within the subject and within narration itself, in time. And, finally, the Lacanian text differs from the deconstructionist (problematic) text, to which it bears a resemblance in that the "positions" of narration ("I" and "you," for example) are anchored, held down, by Freudian desire—the motor and object of Lacanian cognition—which enforces a curb on the infinite textual regress opened by *différance*. The "signifier"—subject to desire, as Lacan states—"stops the otherwise endless movement (glissement) of the signification" (Lacan 1977b: 303). Hence the Lacanian text—to Derrida's outraged dismay (see "The Purveyor of Truth" in *The Post Card* [1987])—may speak its own "truth," the Other's "truth," what Jacqueline Rose calls "the truth of the unconscious [that] is only ever that moment of fundamental division through which the subject entered into language and sexuality, and the constant failing of position within both" (1982: 53). Thus, there is no knowable subject in Lacan's text or in Derrida's (no Other of the Other). Lacanian psychoanalysis and deconstruction do not differ here. But the desire of the Other—a cultural institution not integrally part of the Derridean program—positions and limits the free play of signification through the continual resubjection of the signifier (the subject) to the Other's desire, through the continual "passage into the semiotic triangle of Oedipus" (Spivak 1977: 222).

Difficult Cognition

One can scarcely examine the Lacanian cognitive enterprise without being mindful of the famous "difficulty" of reading Lacan's own discourse. Most scandalizing about Lacan's discourse is its own disruption of clear apprehension in its self-division, the fact that his own "explanations" seem self-contradictory, *split*. For Lacan, the pervasive figure of the split indicates a fundamental division in psychic life, in selfhood, and even within the things we know, within cognition itself. In literary studies, it is a permanent division within the text and narration; in cognitive studies it is a permanent incongruence between phenomena and understanding—truth and meaning—the absence of the "reasonable harmony" between them upon which Whitehead establishes the truth of cognition. Lacan's model of discourse, accordingly, is not that of a unified thing but of a split process, a two-fold process that swings metronome-like from side to side between product and production (manifest text and unconscious discourse), back and forth, and never reaches a point of stability or wholeness. This narrative model poses a serious threat to the empirically-based tradition of interpretation as a transparent and focusable lens, an open subjectivity, through which a detached investigator peers into a stable (possibly pictographic) narrative structure. As we have seen, in this complicated undecidability narrative understanding undermines the simplifying procedures of empirical studies of cognition.

It is not surprising that the threat of Lacan's "return to Freud"—a return to concern with the unconscious system that engenders psychic life and virtually makes subject and cognition possible in the first place—has met tremendous resistance in practice and theory in the United States and England, more than was met by deconstruction's emergence during the same period and certainly more than was met by the emergence of cognitive science—the science of "mind"—in the 1950s and 1960s. A well-informed critic such as Edward Said, for example, once worried that projects such as Lacan's "return to Freud" will only establish "a [new] canon whose legitimacy is maintained with loyal devotion." "A new canon," he continues, "means also a new past or a new history and, less happily, a new parochialism" (1983: 143). Michael Ryan once announced that Lacan is only superficially a "maverick" in psychoanalysis and is really a "clever fundamentalist, rather conservative, clearly antimarxist, roundly antifeminist, and theocratic" (1982: 104). Curious, here, is the sharp edge—the near frenzy—of such resistance in the face of strong evidence to the contrary. The influence of Lacan's critique of the subject now runs quite deep in the very areas mentioned, in Marxist, feminist, and deconstructive criticism. Juliet Mitchell suggested in 1974 that psychoanalysis in particular tends to generate such opposition: "though the criticism *seems* to be over specific issues," its aim is the "whole intellectual framework of psychoanalysis," particularly the dimension of the unconscious that Lacan has insisted on so fervently (1974: 5). Mitchell's suggestion recalls Freud's notion that psychoanalysis takes hold theoretically most firmly where it is initially most strongly resisted. Such resistance, Freud implies, is a sign of unconscious recognition and a defense—simultaneously an expression and a disavowal of desire. This process of acceptance appears to be taking place in the United States and England.

Positioned in a different paradigm from that of cognitive science, Lacan's concept of narration and cognition rests squarely on an ontological fault line, the radical split of a subject irretrievably unwhole—the subject of what Lacan calls *aphanisis*. Lacanian analysis, in this regard, is difficult because it is revolutionary in its promotion of this paradigm of cognition. Lacan's greatest difficulty, then, is precisely this: because the subject, for Lacan, is marked by this irrevocable split, what we are accustomed to calling unity and wholeness in form and seeing as concepts centrally important to narration and cognition are unceremoniously ousted. But these concepts do not just vanish; we still have some reason to speak of wholeness and unity. Such concepts are relegated, however, to the status of being (in Jacques-Alain Miller's term) a mere "suturing" over of the fundamental split with the various commitments (threads) of ideology, the inevitable cultural bias that we bring to any one approach to the knowing subject in narration in hopes of promoting a view of meaningful significance and wholeness. In shifts such as these—in which sense and nonsense, the central and the marginal, are seen to switch places—we stand witness as the lines of understanding palpably move from one paradigm to the other, from a world in which they (already) made sense to a world in which they are (just now) making sense. This revolution of the knowing subject's status, and the resultant shift in the way we understand narration, poses another difficulty. This difficulty, more than any other, focuses on the curious relationship between understanding and desire, a relationship parallel to that between cognition and narration.

Lacan's Allegory

The split in the subject and in narration we are describing takes its most palpable form in the allegory of Lacan's own discourse. In "The Freudian Thing," Lacan attempts to present "the return to Freud in psychoanalysis." The form the presentation takes is an allegorical narrative, a kind of polemical ventriloquism, in which Lacan speaks for Truth herself, for his adversaries, and in a section entitled "the discourse of the Other," for the desk behind which he stands as he talks. In this essay the narration of his discourse on and with *things* takes the forms of a colloquy with his audience (riveted "so respectfully in those seats to listen to me despite the ballet of calls to work" [1977b: 132]), the plot of a murder mystery (1977b: 123), the frenzy of a tragic chorus (1977b: 124), the nonsense of a joke (1977b: 122), the allegory of Acteon's dismemberment (1977: 124), a "game for four players" (1977b: 139), and the discourse of the desk (1977b: 136).

For Lacan, things are problematic: Truth herself says, speaking through Lacan, that "the trade route of truth no longer passes through thought: strange to say, it now seems to pass through things. . . . Here, no doubt, things are my signs, but, I repeat, signs of my speech" (1977b: 122–23). As opposed to this truth, Lacan defines the discourse of the desk as "piling pleonasm on to an antonomasis" (1977b: 136)—that is, multiplying language on a discourse that already misnames its object. The discourse of "Truth" and the definitions of the speaking desk present a confused conception of "things" by making them both the objects and the signs of cognition. The discourses—that of "Truth," and that of a supposedly dumb desk—bring together things and signs, truth and meaning. Allegory brings together things and signs in a similar way. In allegory, meaning is articulated by means of things (like a desk or like the elements of Freud's dream), and this very articulation, as Lacan says, can signify speech. That is, the very *enactment* of speech in allegory itself signifies the "truth" of an Other speech—which, as Joel Fineman has noted, is "a direct translation of the etymology of allegory: *allos*, other; *agoreuein*, to speak" designates the "discourse of the Other" (1981: 46). For this reason, allegory is, as Fineman says, "both theme and structuring principle" in psychoanalysis (1981: 27), both the theme and the structuring principle of psychoanalytic cognition.

The central question for Fineman, following Lacan, is the central question of the relationship between narration and cognition, between experience and meaning: how does theme or structure manifest itself in time? How is the true "reality" of time transformed into the meaningful "idea" of temporality? How can the polyphony of ventriloquism—the *split* of discourse we discussed earlier—surmount its own tumult? "In order to appreciate the scope of this split," Lacan says in "The Freudian Thing,"

> we must hear the irrepressible cries that arise from the best as well as the worst, attempting to bring them back to the beginning of the chase, with the words that truth has given us as viaticum: 'I speak,' adding: 'There is no other speech but language.' The rest is drowned in their tumult.
>
> 'Logomachia!' goes the strophe on one side. 'What are you doing with the preverbal, gesture and mime, tone, the tune of a song . . . ?' To which others no less animated give the antistrophe: 'Everything is language: language when my heart beats faster when I'm in a funk, and if my patient flinches at the throbbing of an aeroplane at its zenith it is a way of saying how she remembers the last bomb attack.' (1977b: 124)

In Lacan, as in Whitehead, meaning is everywhere. But unlike Whitehead's contention, this "meaning" is not the truth of things but the phenomena of understanding—which is structured like a language. Things have a "way of *saying*," the way of allegory by which cognitive truth comes to inhabit meaning.

Paul de Man, following Walter Benjamin and others, has attempted to understand the mechanisms of allegorical cognition—the structure of allegory—in antiromantic terms. His discussion can help define the element of unconscious desire inhabiting cognition. Allegory, he argued, eschewed the nostalgia enacted in Coleridge's "symbol" for the more difficult pleasures of understanding and self-knowledge. The knowledge of allegory, as de Man describes it, is not conscious and articulate but inhabits the structure of allegory. It is the knowledge of the split between truth and meaning enacted in the very split between narrative and significance that enables allegory altogether. In allegory, he argues, we have "a relationship between signs in which the reference to their respective meanings has become of secondary importance" (1969: 190). Such a relationship reduces meanings to signs in the same way cognition reduces experience to apprehensible relationships. Geoffrey Hartman also points to this *cognitive* aspect of allegory when he asks whether there can be "nonallegorical kinds of reading" (1980: 88–90). By "nonallegorical" readings Hartman means something akin to Blake's visionary rather than his allegorical poetry, the possibility of pure and immediate cognition (like the pure and immediate cognition Whitehead sees in mathematics) rather than cognition mediated through structures of signification such as we examined in Greimas's narratology or even Darwinian structures of selection. For Hartman, allegory, like cognitive structures more generally, "subdues" texts through a kind of reductive mediation. Like cognitive structures, allegory, "though driven by a desire for transcendence, remains skeletal, grimacing, schematic; . . . the no man's land between what can and cannot be represented" (1980: 90).

In psychoanalytic terms, then, allegory functions by means of repression and, more specifically, by means of making its own cognitive functioning "unconscious." Psychoanalytic terms are useful because they make clear the elements of power and of desire within cognitive activity. They make it possible to conceive of "unconscious" cultural institutions inhabiting cognition. In this conception of unconscious cognition, the functioning of binary structures in the generation of meaning is like the function of the Father in Freud: "the essential process behind the so-called paternal metaphor," Régis Durand has written, "is a repressive gesture, the constitutive repression: a passage from a world of pure difference and meaningless oscillation to an anchoring, a stabilization through some key symbols" (1981: 50). The father is not "himself the author of the law," André Bleikasten has written. Rather, he owes his authority "to the specific place which he comes to occupy within the family configuration in relation to the mother and child—a place . . . 'marked in life by that which belongs to another order than life, that is to say, by tokens of recognition, by names'" (1981: 199).

The "paternal metaphor," what Lacan calls "the name-of-the father," is a version of psychoanalytic allegory that can be mapped on a Greimassian square.

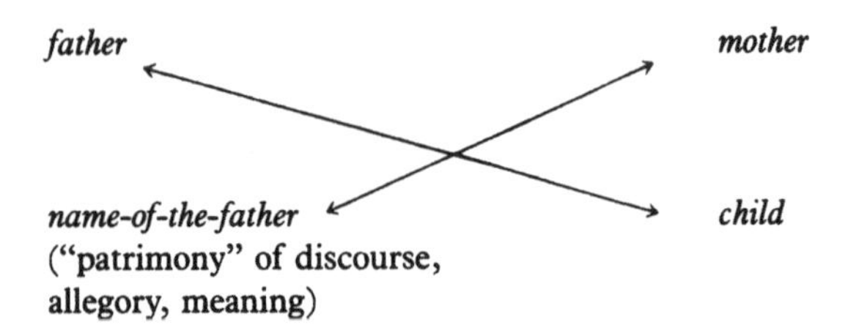

In this square, the name-of-the-father is precisely the allegorical and cultural meaning of biological facts. It is the allegory of that other order than life, the discourse of the Other. That "Other" is the discourse of cognition—linear, temporal, coherent—which requires precisely the repression of the difference between truth and meaning, the difference Lacan describes between the "Real" and the various signifying systems—the Symbolic, the Imaginary, the Symptom—he aligns against the Real.

These systems also can be inscribed in a Greimassian square, though they exist at a higher level of abstraction than the family square we just presented: their superimposition on that square suggests provocative equations between actors in the family romance and culturally specific modes of signification.

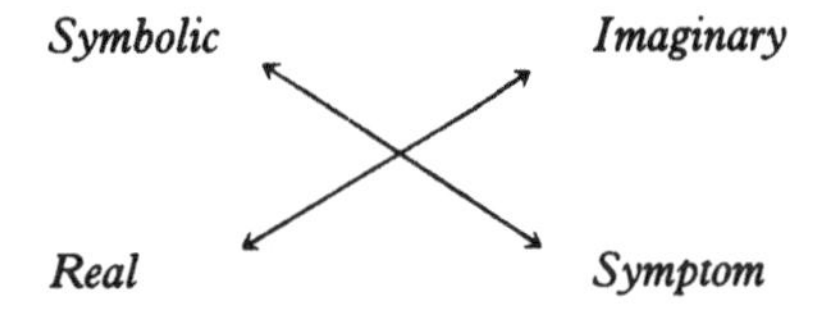

The opposition between the Symbolic and the Imaginary orders in Lacan contrasts two modes of understanding. The Symbolic mode, Durand notes, is "the successful achievement of the 'family romance,' a linear noncontradictory discursive and symbolic mode"; and the Imaginary mode is "opposed to linear coherence, emphasizing discontinuity, oscillation, and nondifferentiation" (1981: 51). The third element, the Symptom, presupposes the Imaginary in its *enactment* of meaning within bodily life. It is, in fact, a kind of *literal* or *empirical* allegory, inscribing meaning in things. Yet it also underlines the discursive order of the first level in its opposition to the Real. In this schema Lacan, like de Man, distinguishes between psychoanalytic allegory and other modes of understanding. Whereas propositional discourse presents statements about the world which can more or less indifferently be described in terms of truth or meaning, allegory "explodes" understanding into the opposition between truth and meaning by creating gaps or the illusion of gaps in the continuity of the all-encompassing ecology of cognitive relationships. It does so by marking or, in Lacan's term, "punctuating" understanding, emphasizing the opposition of the "Real" (in contrary and contradictory ways) to different modes—culturally determined modes—of the cognitive apprehension of the real. Most important in Lacan is that the subject is and *has to be* quite literally unconscious of the Real. Whitehead suggests that the absence of logical relationships between things is unspeakable and unthinkable in an argument for the cognitive apprehension of the "key" to reality. Lacan, also arguing the unthinkability of nonrelationships, suggests the "Real" is outside cognition.

It is outside cognition for Lacan, but not without effects on cognition. The reason for this is that the "punctuation" Lacan describes is governed by desire, which he situates between the "need"—inhabiting the "real," physiological world of the infant—and "demand," the symbolic social articulation of discourse. It is in terms of desire that we can more precisely define Lacanian cognition, Lacanian allegory. "Allegory," Fineman argues, "initiates and continually revivifies its own desire, a desire born of its own structurality. Every metaphor is always a little metonymic because in order to have metaphor there must be a structure, and where there is a structure there is already piety and nostalgia for the lost origin through which structure is thought. Every metaphor is a metonymy of its own origin, its structure thrust into time by its very structurality" (1981: 44). Allegory, then, is *temporizing* in both senses of the word: both *passing* and *gaining* time so that the endless displacements and endless destructions of time (of the Real) can be deferred. "Truth," "strophe and antistrophe," the desk, the Freudian "thing,"—each recognizes in its name and structurality, as allegory does, what de Man calls "temporal difference" that is the structuring principle of allegory (1969: 191) by combining arbitrary names with the activity of cognition. The "temporality" of allegory, then, is the site of what we call in Chapter 7 the *emergence* of meaning in narrative. It marks the complexity of the "temporal difference" between that enactment of meaning and the "lost origin" of truth to which that meaning refers. In this complexity, narrative temporality itself is not a pole in a binary opposition, but arises out of the oppositions between complex "Symbolic" significance and binary "Imaginary" significance (on the level of meaning) and between reductive allegorical "Symptoms" and the undifferentiated "Real" (on the level of truth).

Such temporality is the condition of desire. As Peter Brooks notes, "desire has a history, the story of its own past, unavailable to the conscious subject but persistently repeating itself in the transference, in the symptom" (1979: 78). Symptom, like the imaginary, is the opposite of narrative as we have described it. It is the nontemporal identification of truth and meaning. As such, it is the story not of the absent but of the phantom father, a ghostly "presence" that displaces the subject in the failure of repression. "The original repression," Durand writes, "is not so much rejected as grotesquely twisted, played with, and displaced" (1981: 62). Desire, on the other hand, is not a symptom but a temporal activity that opposes the "truth" of reality with the meaning of narrative. The Real, as Lacan says of perversion, is "outside of time": "The fantasy of perversion is nameable. It is in space. It suspends an essential relationship. It is not atemporal but rather outside of time. In neurosis, on the contrary, the very basis of the relationships of subject to object on the fantasy level is the relationship of the subject to time" (1977a: 17). Lacan opposes a "thing" to an "object" here because "a real thing, one that has not yet been made a symbol, . . . has the potential of becoming one" (1977a: 46). The neurotic lives in a world of objects where everything is meaningfully (if confusedly) inhabited by the temporality of desire—in "neurotic behavior," Lacan writes, ". . . the subject tries to find his sense of time in his object, and it is even in the object that he will learn to tell time" (1977a: 17)—while the psychotic lives in a world of things and "disavows" time altogether.

Lacan's allegory, then, attempts to address both worlds, to bring together things and objects, the world in which "things" simply resist understanding and the world in which "objects" are understood, cognitively apprehended—the "good story" and the "well-formed argument" of Bruner's cognitive topology. "Of all the undertakings that have been proposed in this century," Lacan writes, "that of the psychoanalyst is perhaps the loftiest, because the undertaking of the psychoanalyst acts in our time as a mediator between the man of care and the subject of absolute knowledge" (1977b: 105). The psychoanalyst is the mediator of knowledge, of cognition, his medium is language, and his rhetoric is an allegory of desire that transforms binarity into a spacious dialogue, the dyad of the analytical situation into a "game for four players" (1977: 140). He does so by articulating in his very silences the discourse of the Other to initiate a dialogue of desire.

Knowledge and Power

The opposition between truth and meaning Lacan presents is parallel to the vocative and accusative languages in the rhetoric of old people we examine in the next chapter as the discursive and cultural elaboration—the "special case"—of the empirical studies we surveyed in Chapter 3. In fact, this opposition can help us to situate the intersection of subject and culture more fully than the empirical paradigm that reduces the individual to the passive "subject" of overwhelming transcendental forces of what is naturally "given." If, as Latour argues, "truth" is underwritten by social and historical activities, and if, as we have argued, natural selection is modified by "artificial" selection (as embodied in the negativity of "weak" preparedness), the relation of the subject to truth is more complicated than empiricism suggests. That complication is the "inscribed" opposition (to use Latour's figure) between truth and meaning.

Lacan inscribes this opposition in his allegory, but Greimas elaborates it semiotically in *Structural Semantics* when he presents two modes of semantic analysis, qualificative analysis that analyzes the subjects of discourse as discrete units, as active subjects and agents, and functional analysis that analyzes the subjects of discourse as constituted by their action, as passive subjects of knowledge (see 1983b: 139–40). The first is the situated subject of aging cognition we presented in Chapter 3 in which the activity of the (individual) sender is of utmost importance. The second is the passive subject of actantial modes and genres we described in Chapter 2 in which positions and roles in which actors find themselves are most important and which culminates in the "passive" activity of reception. Like Freud's psychosis and Lacan's perversion, like speech-act theory—including Lacan's own radical speech-act theory which asserts "even if it communicates nothing, the discourse represents the existence of communication; even if it denies the evidence, it affirms that speech constitutes truth; even if it is intended to deceive, the discourse speculates on faith in testimony" (1977b: 43)—the first mode of semantic analysis posits, as Freud says of psychosis, a conflict between the ego and the outer world. The second form of analysis, "functional analysis," like neurosis, like systematic structural cognitive analysis, constitutes the subject in relation to its activities—and the logical configuration of possible activities—by positing, as Freud says of neurosis, a conflict between the ego and the id (Freud 1963: 185).

Whereas semiotic and structural analyses usually take the form of actantial analysis, assuming as they do that the *positions* of analysis, usually analogous to the positions of grammatical analysis, yield the simplest and most logical understandings, Anglo-American cognitive science usually takes the form of "functional" analysis (like that of Propp) and assumes that the analysis of activities (empirical behavior) rather than actants yields the most empirical understandings. In *Structural Semantics* Greimas acknowledges the existence of both kinds of analysis. "Paraphrasing Lacan," he writes, "we can say that two kinds of madness await mankind: on the one side, schizophrenia, the exaltation of total freedom in communication, ending in non-communication; on the other side, a completely socialized and iterative speech, Queneau's 'You talk, you talk, that's all you know how to do,' which is also the negation of communication, discourse deprived of information" (1983b: 39). The former tends toward what we might describe as "infinite semiosis," while the latter tends toward the multiple and scattered details of empiricism, a kind of unfocused (and nonnarrative) garrulousness occasionally associated with stereotypical old age.

Greimas is describing the radically "bi-isotopic" nature of language inscribed in the distinction, of great importance to Lacan, between "enunciation" and "statement" (*énoncé* [see Schleifer 1990: 180–87]). The subject of "enunciation" selects the elements of discourse, performing an act that presupposes and "enacts" that subjectivity ("whatever . . . any enunciation speaks of," says Lacan, "belongs to desire" [1978: 141]), whereas the subject of "statement" is "determined retroactively" (1978: 139) by means of a functional analysis of discourse, an analysis of the subject that its combination of elements implies. As we have seen, Greimas understands linguistic activity as first a "series of messages" upon which a "systematic structure—the distribution of roles to the actants—is superimposed" so that it becomes "a simulator of a world from which the sender and the received of a communication are excluded" (1983b: 134). In Chapter 2 we examined the "structure" of actants and the cognitive awareness to which they give rise, including the recognition of the "external" form of discourse defined in relation to the "receiver" of messages. By examining Lacanian psychoanalysis, however, we can retraverse this territory not from the point of view of the conscious subject of knowledge who experiences discourse, and not even from the viewpoint of the *social* subject of knowledge situated within a community or "sociobiological group" of species activity, but as the subject of discursive knowledge who articulates a language that is both within his or her possession and beyond it—a cognitive subject who is both conscious and unconscious. For this subject there is the particular time of the enunciation—best represented, perhaps, by Lacan's famous variable psychoanalytic sessions—which is a "thing" that, like all "things" in the Real, is a horizon only, with only the "potential" of becoming a symbol. More important, there is another kind of time, that of the statement represented in the tenses of discourse, which can only be symbolic. The time of enunciation is, as Lacan says, "outside of time," and it becomes temporal only when it is structured and "symbolized" in discourse.

Such discourse is the allegorical narrative we have been examining. It is the intersection of qualities and functions, Real and Symbolic time—the stories and logic of cognition. Narration, in Greimas's succinct formulation, creates a temporal succession "which is neither pure contiguity nor a logical implication" (1983b: 244), neither the contiguity of enunciation's socialized and iterative speech (the very repetitive speech of old people) nor the necessary succession of the statement's systematic structure. This negative opposition ("neither . . . nor") is best visualized in the optical illusion, the outlined cube, for instance, whose forward side can also be seen as its bottom, but never both at the same time. The two cognitive activities of such an optical illusion are in a relationship to each other that is neither logical implication nor pure contiguity. In *The Confidence-Man* one of Melville's characters tries to explain such a picture: "If the bill is good, it must have in one corner, mixed in with the vignette, the figure of a goose, very small, indeed, all but microscopic; and, for added precaution, like the figure of Napoleon outlined by a tree, not observable, even if magnified, unless the attention is directed to it" (1967: 346). Napoleon or Melville's goose is *there*, but it is figured by something else. This figuring, however, is not a form of metaphoric substitution or even the semantic transference Ricoeur and Davidson argue is the function of metaphoric discourse, but is accomplished by means of metonymic spacing. Yet the metonym *functions* like substitution, like a metaphor. We cannot see both Napoleon and the tree at the same time, but only one or the other.

The same "optical illusion" effect is inscribed in what Greimas calls "linguistic activity," and more generally it is inscribed in cognitive activity. Discourse, as we have suggested throughout this book, offers two modes of cognition. It is essentially split between what Greimas describes as the "power" of enunciation and the "knowledge" of statement. To articulate this split, Greimas adds *desire* to the relations of knowledge and power in cognition (1983b: 202; see Schleifer 1987a: 184–90). In this way, like

Lacan, he inscribes desire within a framework of cognition that makes the seeming paradox of "instituted" truth we have mentioned throughout this chapter at least conceivable. The juxtaposition (or superimposition) of Lacan and Greimas makes this evident. In Lacan the sender *complexly* embodies the cultural values, what it sends is the discourse of the Other, so that the level of knowledge—the level of cognition—creates the effect of a "subject" of discourse as well as cultural values. Here, then, is the *split* in knowledge which Lacan enacts and to which he continually returns.

Moreover, it is precisely because knowledge is always another's knowledge, precisely because knowledge itself has a history and emerges in time and in relation to a society of human subjects, that *instituted* cognition is possible. This is what Latour means by the "mobilization" of cognition: knowledge always arises in "agonistic encounters" (1986: 5) that are always, at least on some level, interpersonal. What the "structure of the signifying chain discloses," Lacan writes, "is the possibility I have, precisely in so far as I have this language in common with other subjects, that is to say, in so far as it exists as a language, to use it in order to signify *something quite other* than what it says" (1977b: 155). The *difference* between the messages of knowledge and power—between cognition and affect—arises in the fact that there is always more than one subject of discourse, more than one subject of knowledge. Because of the existence of *more than one*, it is not the difference of a binary opposition but rather an enabling difference, like the optical illusion's double vision, that creates power and knowledge (the "truth" of affect and the "meaning" of signification) that are not symmetrically reducible to one or the other. This very irreducibility realizes itself in the phenomenon of desire. Desire, Lacan says, is the desire of the Other. The object of psychoanalytic transference, the "subject supposed to know" (*sujet supposé savoir*), *is* supposed to know, Lacan says, "simply by virtue of being the subject of desire" (1978: 253). That is, the knowing subject is both the object of the patient's desire and the subject of desire itself. And desire itself is both conscious and unconscious: it arises in the confrontation between the irreducible difference between conscious demand and unconscious need situated in a world of other people.

The very "structurality" of allegory is this difference (embodied in desire) between power and knowledge, and in this difference the effect of temporality arises, that "sense" or feeling of time we have yet cannot reduce to cognitive articulation. "What, then, is time?" Augustine asks. "If no one asks me I know; if I want to explain it to someone who does ask me, I do not know" (1960: 287). Time, like conscious-unconscious desire, exists between the *power* of its felt presence—the "truth" of its affect—and our articulated *knowledge* of it—the "meaning" of its effect—in which its experience never quite corresponds to its "understanding." Discourse "divides time," as Hartman says, "to make us aware of it" in what he calls "a ghostly, dimensional shift from action to observation" (1975: 290, 287; see also Ricoeur 1984, 1985, 1988), from power to knowledge, from the discourse of the self to the discourse of the Other. To paraphrase Augustine, we are sensitive to the punctuations, the differences, of time, yet those punctuations cannot be named because they are the ground of naming and articulation altogether, the ground of the experience of cognition into which (in another ineluctably temporal passage) we are born. For this reason, time is closely connected to desire, and both are forms of cultural cognition. In Augustine's answer to "What is time?" temporality inhabits the desire to explain engendered by the separate existences of others who are neither logical implications of the subject nor (because, as Lacan says, "I have this language in common" with them, a language which is a common cognitive awareness) simply contiguous with the subject.

"Strange Temporality": The Hesitations of Narrative

The paradox of Lacanian understanding is that desire is recognized in the failure of cognition—in the "unconscious," in the slip of the tongue, in the "nonsense" of discourse, in the "impediment, failure, split," as Lacan notes. "In a spoken or a written sentence something stumbles. Freud is attracted by these phenomena, and it is there that he seeks the unconscious. There, something other demands to be realized—which appears as intentional, of course, but of a strange temporality" (1978: 25). Here we have arrived at a kind of theory of narrative meaning that situates it within an ecology of human cognition wider even than those of Greimas and cognitive science we have examined. "The symbolic function," Lacan writes, "presents itself as a double movement within the subject: man makes an object of his action, but only in order to restore to this action in due time its place as a grounding. In this equivocation, operating at every instant, lies the whole process of a function in which action and knowledge alternate" (1977b: 73). This equivocation operates at every instant because cognition is both an activity and the result of that activity, both power and knowledge.

"The author of these lines," Lacan says, "has attempted to demonstrate in the logic of a sophism the temporal sources through which human action, in so far as it orders itself according to the action of the other, finds in the scansion of its hesitations the advent of its certainty" (1977b: 75). The certainty of "objective" knowledge is a function of uncertain activity. The certain knowledge of self with which Descartes articulated the modern ground of cognition is *positioned* in relation to the uncertainty of other subjects. The hesitations Lacan mentions here are what he calls the "dialectical punctuation" of speech effected by the analyst (1977b: 95): "the punctuation, once inserted, fixes the meaning; changing the punctuation renews or upsets it; and a faulty punctuation amounts to a change for the worse" (1977b: 99). Greimas's term for this punctuation is "the intrusion of *history* into *permanence*" (1983b: 293) which "explodes" (1983b: 245) a "complex category" (e.g., "discourse," "cognition") into a "disjunctive category" (e.g., "statement vs. enunciation," "knowledge vs. power"). Literary criticism's term for this punctuation—which can also be understood as "cognition" itself, complexly understood as an instituted activity—is "interpretation."

With this term we return to the opposition between cognition and narrative which we examined in Chapter 1. Lacan describes the place of desire within this opposition. "Let us place ourselves," he says,

> at the two extremes of the analytical experience. The primal repressed is a signifier, and we can always regard what is built on this as constituting the symptom *qua* a scaffolding of signifiers. . . . Although their structure is built up step by step like any edifice, it is nevertheless, in the end, inscribable in synchronic terms.
>
> At the other extreme, there is interpretation. Interpretation concerns the factor of a special temporal structure I have tried to define in the term metonymy. As it draws to its end, interpretation is directed toward desire, with which, in a certain sense, it is identical. Desire, in fact, is interpretation itself. (1978: 176)

The special temporal structure Lacan speaks is the structure of hesitation, a complex figure, a complication of the literal and the figurative. Literally, it is a pause, a stopping place, a kind of Wordsworthian "spot of time." Yet insofar as the time of enunciation is relentless, it is only a figure for something else, the illusion of stopping, of meaning, of the momentary apprehensions of cognition. Hesitation is marked by difference that can always be "known" by means of the logicoscientific differences of binary oppositions—the typology of binarity inscribed in the semiotic square—yet whose phenomenal effects, like the time Augustine describes, never quite correspond to the logic of binarity.

That is, the "hesitations" of time and meaning, like the logic of binarity itself, only exist in a system of differences that is a cultural, not a logical, artifact. They exist, as Lacan says, in the manner of the call slip in the library: "It is the realist's imbecility," he writes, "which does not pause to observe that nothing, however deep in the bowels of the earth a hand may seek to ensconce it, will ever be hidden there, since another hand can always retrieve it, and that what is hidden is never but what is *missing from its place,* as the call slip puts it when speaking of a volume lost in the library. And even if the book be on an a adjacent shelf or in the next slot, it would be hidden there, however visibly it may appear. For it can *literally* be said that something is missing from its place only of what can change it: the symbolic" (1972: 55). The hesitations of interpretation, like cognition itself, only exist within a framework of cultural meanings and cognition in which they must occur. These hesitations are analogous to what Lacan calls "an indirect discourse, isolated in quotation marks within the thread of narration, and, if the discourse is played out, it is on a stage implying the presence not only of the chorus, but also of spectators" (1977b: 47). "The discourse in an analytic session," he says elsewhere, "is valuable only in so far as it stumbles or is interrupted" (1977b: 299). What it stumbles over, what it is interrupted by, is the space of desire, the dialogue or superimpositions of one discourse and another.

That is, the special temporal structure creates and fills gaps and responds to nothing—to Hamlet's ghost, to the uncanny doubleness of language that Lacan, Greimas, and, as we will see in the next chapter, old people describe, to the Other's desire—the nothing that is not there and the nothing that is. In "The Lady with a Dog" Anton Chekhov describes what we are calling "instituted" cognitive activity that brings together knowledge, power, and desire in a love story. Gurov, a middle-aged visitor to Yalta, accustomed to casual love affairs with various women, meets a young married lady with a dog and begins an adulterous affair with her. He is somewhat bored with the intensity of her feeling and guilt until the day they go to the mountains. Here is the only place in the story that the otherwise impersonal and omniscient narrator—the subject of knowledge—interrupts the continuity of the story with an explicit enunciation that betrays and exceeds the story's illusion of objectivity.

> Not a leaf stirred, the grasshoppers chirruped, and the monotonous hollow roar of the sea came up to them, speaking of peace, of the eternal sleep lying in wait for us all. The sea had roared like this long before there was any Yalta or Oreanda, it was roaring now, and it would go on roaring, just as indif-

> ferently and hollowly, when we have passed away. And it may be that in this continuity, this utter indifference to life and death, lies the secret of our ultimate salvation, of the stream of life on our planet, and of its never-ceasing movement toward perfection.
>
> Side by side with a young woman, who looked so exquisite in the early light, soothed and enchanted by the sight of all this magical beauty—sea, mountains, clouds and the vast expanse of sky—Gurov told himself that, when you came to think of it, everything in the world is beautiful, really. (1979: 226)

Such a lyrical passage presents the allegory of knowledge: the roar of the sea equals "continuity" equals the "never-ceasing movement toward perfection" of time equals "salvation." This allegory is the irruption of enunciation into the story's statement. It is the "shifting in" from the objectivity of the story to the narrator's first-person act of enunciation.

As such, it articulates what is literally *unconscious* to the characters—the discourse of the Other/sender—which, nevertheless, creates a framework that conditions the vague feelings they have about their experience (which, like the experience of time, they cannot *speak*). Moreover, even the propositions enunciated in this passage by the narrator are articulated against—are punctuated by—"unconscious" binary opposition whose second term—"discontinuity" or "movement toward annihilation" or "neither salvation nor perdition, but simply indifference"—cannot be articulated within the cognitive framework of this love story. By allegorizing experience, Chekhov "immediately" erases the temporal constraints that condition that allegorization—including the temporality implicit in hierarchical binarity itself—to achieve the impersonality of "salvation" and system. After the meditation on the indifferent sea, the realization of *meaningful* time, the whole world seems beautiful to Gurov; everything is somehow transformed. This is the narration of cognition, which takes place in time but erases temporal constraints in the face of "meaning."

At the end of the story Gurov repeats this cognitive act. He sees himself in a mirror, his hair turning gray. The momentary interruption of his vision of himself, the hesitation in his narrative, transforms his intention "to fondle her with light words" into the narration of his love life culminating in the realization that "only now, when he was gray-haired, had he fallen in love properly, thoroughly, for the first time in his life" (1979: 234). Here is a kind of "mirror stage" creating a hesitation in the time of narration. In Gurov's gaze, as Lacan says, the object is "lost and suddenly refound in the conflagration of shame, by the introduction of the other. Up to this point, what is the subject trying to see? What is he trying to see, make no mistake, is the object as absence" (1978: 182). The object in absence is both Greimas's and Lacan's object of desire, and more than this, it is the object of knowledge, of cognition, altogether. It is the apprehension of experience *as* meaningful, the *unconscious* act of cognition which seems, to the knowing subject, to be simply perception. Gurov's "proper" love arises not with the presence of this object of desire and cognition but with the allegorical recognition, marked by his gray hair, of temporality and death—with its absence that also carries the possibility of the dialogue Lacan speaks of. That is, the discovery of beauty and love, of "ultimate salvation," is what the activity of cognition and the *desire* of cognition offer. Much as the ancient Athenians called the Furies the Eumenides (the benevolent ones) and prayed in Aeschylus's play that they might "Bless them, all here, with silence" (1959: 171), so the desire of cognition articulates—it falsely names—the horror of indifference as "salvation." Gurov, in effect, names the intimations of his own mortality that he sees in the mirror as "love."

Such false namings, such an allegory, transformed the Greek Furies into benevolent deities. In Chekhov's story the beauty discovered before the indifferent sea and the indifferent mirror makes the salvation of love possible. It makes salvation possible by discovering and humanizing—in a word, by punctuating—as cognition itself does, the chthonic origins of life, the relentlessness of time, inarticulately expressed in the roar of the sea, the silence of the analyst. In "Narration in the Psychoanalytic Dialogue" Roy Schafer, following Lacan, defines the transference narrative as "new remembering of the past that unconsciously has never become the past" in which "the alleged past must be experienced consciously as a mutual interpenetration of the past and present" (1980: 36). Such interpretation is the punctuation of desire, which is precisely what Gurov enacts, in his allegorical "recognition" before the mirror. At the moment he sees his gray hair in the mirror he intercalates the story of his life: "Women had always believed him different from what he really was, had loved in him not himself but the man their imagination pictured him, a man they had sought for eagerly all their lives. And afterwards when they discovered their mistake, they went on loving him just the same" (1979: 44). Here, in the women loving him just the same, is the representation of experience, the transformation of the time of an indifferent world, in terror and in love, into a human image. It is Hartman's "subduing" uncertain experience with an allegory that betrays time by allegorizing it. This takes place in and by means of the cognitive activity of discourse. Lacan has noted that in psychoanalysis the operative "cure" of hysterics is not memory but verbalization—the patient "has made it pass into the *verbe,* or, more precisely, into the *epos* by which he brings back into present time the origins of his own person" (1977b: 46–47).

Thus, love, in Chekhov's story, inhabits the Symbolic; it is a shadow or unreal fiction, without which the light of life seems unimaginable (see Wilden 1968: 192), and which appears in the hesitations in the narrative, at the moment of cognition and cognitive desire. Chekhov creates love before our eyes, while the meaningless and indifferent "real" hovers underneath, a seeming benevolent Fury, the mother of beauty, an under-roar for his whole story. At the end of "The Lady with a Dog," instead of fondling her, Gurov and Anna sit and discuss their situation: "And it seemed to them that they were within an inch of arriving at a decision, and that a new, beautiful life would begin. And they both realized that the end was still far, far away, and that the hardest, the most complicated part was only beginning" (1979: 235). Chekhov ends by having his characters talk, thereby situating themselves, like the old people we examine in the next chapter, in the complication of the time between beginning and end, in the complication of culture. At the end he—and they—like Lacan, mark time in the story by creating the cultural space for the desire of cognition in which time becomes human, speech dialogical and full, and the subject achieves a little freedom. "I might as well be categorical," Lacan says; "in psychoanalytic anamnesis, it is not a question of reality, but of [meaningful] truth, because the effect of full speech is to reorder past contingencies by conferring on them the sense of necessities to come, such as they are constituted by the little freedom through which the subject makes them present" (1977b: 48). Such meaningful "truth" is the object of knowledge and cognition, and in Lacan it is achieved by situating truth itself within the interhuman milieu of cultural values—within the opposition between sense and nonsense—which are literally perceived in unconscious acts of cognition.

6

THE PLAIN SENSE OF THINGS

Aging and the Rhetoric of Narration

The Situation of Cognition

If, as we suggested in Chapter 3 in terms of empirical sociobiology, the discourse of old people presents the *situation* of language and cognition most starkly, it does so because in the language of old people the seemingly nonphysical "sense" of language and cognition are most closely linked to the material objects that signify and manifest that sense—the sound of words, the physical capacities conditioning cognition, the very breath that carries articulate sound. At no other time are breathing and sense more indistinguishable and yet more distinctly themselves than in the discourse of the elderly. It is here that the global "situation" of cognition—the fact that "meanings," apparently free of time and space, exist and manifest themselves in a world of brute materiality—is most manifest. It is here that the situation of truth in the opposition between sense and nonsense that Lacan describes and that we allegorized in Chekhov's narration of Gurov's gray hair is most distinct. Breathing language captures these situations—it is, we will argue, a situation of violence—by marking the conjunction of body function and sense. Thus, in John Donne's "A Valediction: Forbidding Mourning," breath and sense are absolutely distinct, yet to those around the deathbed they are indistinguishable:

> As virtuous men passe mildly away,
>   And whisper to their soules to goe,
> Whilst some of their sad friends doe say,
>   The breath goes now, and some say, no.
>
> (1967: 44)

Addressing the soul so directly that breath does not matter, virtuous men achieve a transparency of discourse and cognition that seems to be impossible in a world such as that in which the deathbed watchers live, in which the material objects that signify—breath, the tone of voice, the behavior of other people, gray hair—are so easily misunderstood.

The relationship between breath and sense, like that between the existence of the species and the cognitive activity that helps sustain it, is a perennial concern for those interested in the functioning of language. What so fascinates Marlow with Kurtz's dying in Conrad's *Heart of Darkness* is that Kurtz's last words—"The horror! The horror!"—are also his last breath. They are aspirated with the same whisper Donne describes in his virtuous men, yet here their "sense," which Marlow takes to be addressed to himself (if only as part of "the whole universe") and elaborates in such obsessive detail, might only be the sound of breathing.

> I was within a hair's-breadth of the last opportunity for pronouncement, and I found with humiliation that probably I would have nothing to say. This is the reason why I affirm that Kurtz was a remarkable man. He had something to say. He said it. Since I had peeped over the edge myself, I understand better the meaning of his stare, that could not see the flame of the candle, but was wide enough to embrace the whole universe, piercing enough to penetrate all the hearts that beat in the darkness. He had summed up—he had judged. "The horror!" He was a remarkable man. After all, this was the expression of some sort of belief; it had candor, it had conviction, it had a vibrating note of revolt in its whisper, it had the appalling face of a glimpsed truth—the strange commingling of desire and hate. (Conrad 1971: 72)

Marlow describes in Kurtz's last words the commingling of desire and hatred, the conjunction of discursive summing up and pure, "vocative" nonsense. Unlike the "pure" cognitive sense disjoined from breath in Donne's poem, Kurtz's words are whispered to the soul *and* to the universe. They are the junction of the self and the earth Wallace Stevens describes in his late poem, "The Plain Sense of Things," as "inevitable knowledge, / Required, as a necessity requires" (1954: 503), the conjunction of self and body Yeats describes in "The Tower," "decrepit age that has been tied to me / As to a dog's tail" (1971: 409).

Kathleen Woodward, in her study of the late poems of Eliot, Pound, Stevens, and Williams, argues that old age offers aged poets "models of wholeness" that are "above all . . . characterized by humility . . . that contradicts the Western way of thinking about mankind in the world—imperialism over nature and other people" (1980: 8, 15). She argues that aging *positions* the elder to see "the whole of the system" (1980: 168). "The wisdom of these four poets inheres in their having recognized, and accepted, the 'other,' what was always the stranger to their work" (1980: 172). She is describing in aged poets what we have already described in old people in general, the *necessity,* born of sociocultural as well as biological causes, to find someone to talk to. Still, if virtuous men are humble, Kurtz is not, and the "humility" before a sense of the "wholeness" of the species and the "otherness" of the receiver who is addressed is not necessarily or easily accepted. As Yeats in old age had said, "nothing can be sole or whole / That has not been rent" (1971: 513). Aging brings the inevitable knowledge of necessities beyond the binary distinctions governing wholeness, otherness, self, and even cognition itself. It brings—it *is*—the violence of the impurity of material change marking ideal conception. As Coward and Ellis write in *Language and Materialism,* "signification is seen (by contemporary semiotic theory) to be a complex heterogeneous process which demands consideration of the extralinguistic" (1977: 134). Age and aging are extralinguistic, yet they meaningfully manifest themselves in the physical sound of voices, the articulations of language, the relationship of listening to hearing—all of which can be figured as the "extralinguistic" phenomenon of breath, the situation of cognition. The rhetoric of old people is perhaps simply cognition, in its combination of knowledge and power, made "plain." Because of the *situation* of old people, this rhetoric offers in its pronouncements and its narrative impulse a sense of the violence embedded in language altogether. Such rhetoric determines (in part) the "roles" of sender and receiver we discussed in Chapters 2 and 3—it defines the "sphere of activity" that Propp and Greimas describe in relation to narrative roles—so that the *emergence* in time of the subjects of experience, their *cultural* institution, is made plain.

Unlike virtuous men, most people find aging ambiguous and impure. What characterizes the elderly is not simply the loss of power, but the situation of simultaneously possessing a sense of self and a sense of otherness about oneself, of understanding "habit," for instance, as both a kind of weary repetition and, as Simone de Beauvoir says, a kind of crystallization in which "the past [is] brought to life again, the future anticipated" (1973: 696). As an eighty-four-year old man told Ronald Blythe in *The View in Winter,*

> Old age doesn't necessarily mean that one is entirely old—*all* old, if you follow me. It doesn't mean that for many people, which is why it is so very difficult. It is complicated by the retention of a lot of one's youth in an old

> body. I tend to look upon other old men as *old men*—and not include myself. It is not vanity; it is just that it is still natural for me to be young in some respects. What is generally assumed to have happened to a man in his eighties has not happened to me. . . . Yet I resent it all in some ways, this being very old, yes, I resent it. (1979: 185)

Aging combines one's ideal sense of oneself with the inevitable fact of one's own materiality—it combines sense and breath in a way that insists, as violence does, on not being ignored. Thus, the old man Blythe is interviewing goes on to say, "King Lear said, 'When the mind's free the body's delicate,' and that is true. . . . I feel so alive, but my muscles tell me otherwise" (1979: 186).

Rhetoric as Violence

Throughout his work Jacques Derrida focuses on the relationship between language and cognition in a way that emphasizes what he calls "the original violence of discourse" (1978: 133). Greimas also notes this violent aspect of language when he describes the generation of the binary oppositions of cognition as an "explosion" of meaning (1983b: 144) and an explosion of the semiotic square (1988: 45), but he hardly emphasizes the violent aspect of meaning and cognition. Derrida, on the other hand, makes the violence in cognition, and especially in its discursive formations, the object of his attention. In an important way, this makes him "poststructural" to Greimas's structuralist sense of meaning and cognition. The violence of language, Derrida argues, is a function of the self-contradictory senses of "writing" that inhabit language: "writing in the common sense is the dead letter, it is the carrier of death. It exhausts life. On the other hand, . . . writing in the metaphoric sense, natural, divine, and living writing, is venerated" (1976: 17). "Natural writing," he says, "is immediately united to the voice and to breath" (1976: 17). In it the accidental materiality of language—the acoustical image, the marks on the page, the *sound* of breathing—becomes a transparency for thought.

But the violence embedded in language is the fact that its materiality cannot disappear: "What writing itself, in its nonphonetic moment, betrays, is life. It menaces at once the breath, the spirit, and history as the spirit's relationship with itself. It is their end, their finitude, their paralysis. Cutting breath short, sterilizing or immobilizing spiritual creation in the repetition of the letter, . . . it is the principle of death and of difference in the becoming of being" (Derrida 1976: 25). Writing cuts breath short by summing up, by judging: with candor, conviction, and the vibrating note of revolt. Last words signify by virtue of their material (and liminal) situation: they do not address the soul, as the "natural writing" of the virtuous man's whispered discourse does in a breathing that is indistinguishable from an ascension to heaven, from the continued possession of "spirit." Rather, their language, like all language—but especially like the rhetoric of old people—addresses the darkness of the universe. It addresses what Derrida calls, following the theologian Emmanuel Levinas, an Other.

Thus, the violence of rhetoric, like that of aging, is the violent conjunction of materiality and spirituality, and its final violence is death. This violence is the fact that spirituality, meaning, and cognition present themselves as simple apprehensions—as self-evident truths—even though they are, in fact, the result of *activities* that can be understood as ("exploded" into) a series of events whose results could be different from the self-evident truths that seem to simply (atemporally) exist. In other words, spirituality, meaning, and cognition can be seen as accidental and contingent, in the same way that aging itself is seen by the old man Blythe interviews (and by the aged Yeats) as accidental and contingent. The materiality of aging is inscribed in the *activity* of language—that is, in *rhetoric*—as that material "Other" which, Derrida argues, is both addressed and invoked (dative and vocative): "the other cannot be what it is, infinitely other, except in finitude and mortality (mine *and* its)" (1978: 114–15). Thus, "the dative or vocative dimension which opens the original direction of language cannot lend itself to inclusion in and modification by the accusative or attributive dimension of the object without violence. Language, therefore, cannot make its own possibility a totality and *include* within itself its own origin or its own end" (1978: 95). In the "accusative" dimension of language, Derrida is describing the cognitive function of language, what J. L. Austin calls the "constative" aspect of language. The constative aspect of language consists of propositions that articulate aspects of the world that can be shown to be true or false. Constative statements, Austin says, are "true or false statements" (1962: 3). Against this view, he posits the "performative" aspect of language, utterances that do not describe things, but enact them. Performative utterances cannot be judged true or false, but rather successful or unsuccessful. Austin notes, for example, that "when I say, before the registrar or altar [in a marriage ceremony], 'I do,' I am not reporting on a marriage: I am indulging in it" (1962: 6). Like Derrida's dative, performative utterances entail the material situation of the activities of language and cognition, which take place at certain times and places. They enact relationships—dative, vocative—between the sender and receiver. In a marriage, it is a relationship between the speaker and the community; in a bet, between the better and his or her interlocutor.

Austin is working in the tradition in Anglo-American philosophy that makes language the object of philosophy, but the binary opposition between constative statements and performative utterances can be seen in the opposition in recent Continental linguistics and discourse theory we have already encountered between *enunciation* and *énoncé* (see Benveniste 1971; Greimas and Courtés 1982). The constative statement articulates the message with its own time (tenses) and place and topological relationships among the actants of the message (nominative, accusative). In fact, as we have seen, Greimas suggests the binary opposition between enunciation and statement (the performative and constative aspects of language) articulates two levels of discourse: that of "power"—a relationship ("dative," "vocative") between sender and receiver—and that of "knowledge," "an objectivizing projection . . . from which the sender and the receiver of a communication are excluded" (1983b: 134). Derrida's "accusative or attributive dimension," the dimension of constative cognition, is a systematic structure. In Lévi-Strauss's terms, it is "an atemporal matrix structure" (1984: 184), whereas the dative or vocative dimension (the dimension of "performance") is a morphemic activity that, above all, calls attention to the materiality of what Derrida calls "the living present."

That materiality is marked by the "otherness" that the old man Blythe interviewed felt in his "own" muscles which did not respond as simply part of himself. Such otherness is very important to Derrida's sense of language, to his sense of the activity of cognition.

> The other is given over in person *as other,* that is, as that which does not reveal itself, as that which cannot be made thematic. I could not possibly speak of the Other, make of the Other a theme, pronounce the Other as object, in the accusative. I can only, I *must* only speak to the other; that is, I must call him in the vocative, which is not a category, a *case* of speech, but, rather the bursting forth, the very raising up of speech. . . . I can speak *of it* only by speaking *to it;* and I *may* reach it only as I *must* reach it. But I must only *reach* it as the inaccessible, the invisible, the intangible. (1978: 103)

The situation of discourse is that of performative enunciation marked against the timeless systematicity of cognitive statement and its tenses. "A *system* is neither finite nor infinite," Derrida says; "a structural totality escapes this alternative in its functioning" (1978: 123). In other words, the very act of cognition violates the Other by inscribing it within the same insofar as it makes it part of the "totality" of a binary opposition.

If this is so, otherness is both addressed and hidden in discourse, in cognition itself. Insofar as cognition and the meanings of language understand phenomena as exemplary of—or through the mechanisms of—larger *systems* of thought; insofar as "the observed entities in the concrete universe," as Alfred North Whitehead argues, "form a particular instance of what falls under our general reasoning" (1967: 22), understanding itself both encompasses and violates otherness. It violates it by inscribing it within the "same." As Derrida says, the violence of language—the activity of cognition—addresses and obscures another "unspeakable" violence, the *incomprehensible* (prelogical, noncognitive) violence of temporality, bodily decrepitude, and death. "Discourse," he writes,

> . . . if it is originally violent, can only *do itself violence*, can only negate itself in order to affirm itself. . . . This secondary war, as the avowal of violence, is the least possible violence, the only way to repress the worst violence, the violence of primitive and prelogical silence, of an unimaginable night . . . which would not even be the opposite of nonviolence: nothingness or pure non-sense. Thus discourse chooses itself violently in opposition to nothingness or pure non-sense, and, in philosophy, against nihilism. (1978: 130)

The nothingness and pure non-sense Derrida describes is "non-sense" precisely because it violates the binarity of Western understanding. It is not even opposite of its opposite. It is precisely this *violence* to opposition—opposing the very "dual periodicity" of breathing and sense Lévi-Strauss (and Donne) describe—that distinguishes Derrida's understanding of cognitive activity from the *systematic* oppositions of structuralism and cognitive science. Moreover, it is this seemingly nonsensical violence that distinguishes the discourse of old people from discourse in general. Discourse in general is the motor of cognition, and as such it eschews breath and imagines that language is transparent and apprehends a reality that is logical (or at least imagines, as Lévi-Strauss does, that logical organization is a property of reality).

In *The Raw and the Cooked* Lévi-Strauss describes discourse in general by describing the locus of the activity of cognition between "the tangible and the intelligible," between perception and logic, in an extended simile with a musical work.

> We can say that music operates according to two grids. One . . . exploits organic rhythms and thus gives relevance to phenomena of discontinuity that would otherwise remain latent and submerged, as it were, in time. The other grid is cultural: it consists of a scale of musical sounds, of which the number and the intervals vary from one culture to another. . . .
>
> The musical emotion springs precisely from the fact that at each moment the composer withholds or adds more or less than the listener anticipates on the basis of a pattern that he thinks he can guess, but that he is incapable of wholly divining because of his subjection to a dual periodicity: that of his respiratory system, which is determined by his individual nature, and that of the scale, which is determined by his training. (1975: 16–17)

The discontinuity Lévi-Strauss describes, figured as breathing, marks the temporality and corporality of "meaning"—the material enunciation of signification and cognition. But in Lévi-Strauss, it does so in a system in which violence is not originary but rather enclosed. Lévi-Strauss is attempting "to transcend the contrast between the tangible and the intelligible" (1975: 14), and he locates signification in the violent contrast between two signifying structures. In doing so, Lévi-Strauss erases the very violence of the contrast he describes, the possibility of violence that extremes of binary opposition create. This is one way of seeing his attempt to dissolve the opposition between form and content in his conception of structure. "Structure," he writes elsewhere, "has no definite content: it is content itself, and the logical organization in which it is arrested is conceived as a property of the real" (1984: 167). (Despite his figure of "explosion," Greimas shares with Lévi-Strauss this attempt to occlude the violence of binarity by conceiving of it as simply a method.)

For Lévi-Strauss, as for Whitehead, the "worst" violence Derrida describes cannot be acknowledged: breath *cannot* be "nothingness or pure non-sense," Kurtz's dying *cannot* not signify, the age of the composer, the idiosyncrasies of his respiratory system, even when they are determined "by his individual nature," cannot inscribe in his work a species of violence that affects the properties of the real. That is, Lévi-Strauss locates both the "organic rhythms" and "cultural" phenomena on "grids" that are signifying structures susceptible to cognition. Like Whitehead, he "locates" a harmony between reason and world. If culture is the positive modification of the latent organic rhythm, then Derrida marks the negative *complexity* of the organic rhythm by inscribing materiality and nonsense in it (in just the same way we marked the complexity of aging cognition in the economy of human life in the Chapter 3). This complexity can be inscribed in Greimas's semiotic square. Greimas's square, as Fredric

Jameson has suggested, maps out what Greimas "takes to be the logical structure of reality itself" (1981: 46), yet it does so, as we have seen, by allowing for the possibility of *narrating* the institution of seemingly self-evident cognitive truths. Using a square, we can see that Derrida "explodes" the negative term "breath" in the binary opposition "sense vs. breath" into its "complex" term, the "materiality of language" (i.e., the material situation of discourse vs. the dead letter):

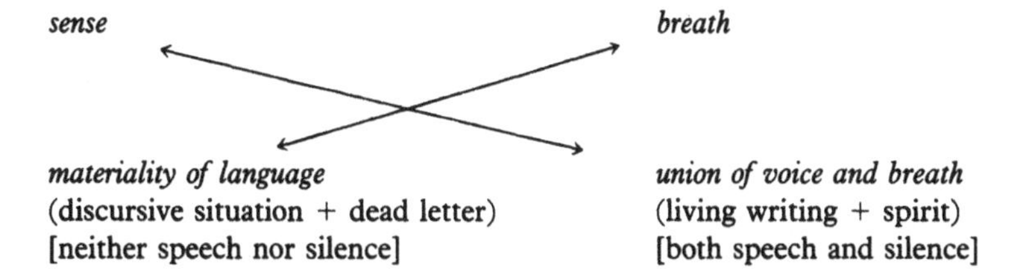

Thus, Derrida argues, discourse in general, like Lévi-Strauss's "system," leaves no room for the dative, performative aspect of language. It leaves no room for the relationship between the sender and the receiver of a message. Discourse in general allows and fosters the illusion of the pure sense of cognition in opposition to the dead letter, and this illusion positions aging as marginal as well as simply liminal, its "impurity" accidental rather than complex.

Aging as Violence

We have commented several times on Lévi-Strauss because, like Whitehead in a different tradition, he offers a strong sense of "pure" cognition by marking the materiality of signification only to erase it in the apprehension of meaning. In this erasure, he presents assumptions underlying an attitude toward aging that sees aging as a monolithic decline of ability and body functions that is simply degenerative, a kind of "accidental" violence in the same way that the accidents of expectation ("more or less") govern Lévi-Strauss's analysis of the signification of music. Thus, traditional approaches to developmental cognitive psychology, have assumed that aging is simply the reduction of behavioral and cognitive effectiveness. Behind this work, however, is the assumption that aging is *simply* and self-evidently degeneration, simply the opposite of the mature culmination of adulthood; that its "meaning" is only a kind of "background" of physiological fact in the same way that Lévi-Strauss offers us the "background" of respiratory "fact" upon which to inscribe the "cultural" grid of musical systems. What it fails to do (as we tried to do in Chapter 3) is to inscribe the significance of the changes of aging *within* its meaning; to make the enunciated violence of aging complicit within its significance. Traditional approaches of cognitive psychology fail to discover within the "pure nonsense" of degeneration (and that worst nonsense of death of which that degeneration is a part) a sense of the *complexity* of human aging. The distinction between "nature" and "culture" upon which Lévi-Strauss bases *The Raw and the Cooked*—a distinction which in the description of music finally privileges the cultural over the natural, sense over breath—is a distinction that has narrowed the examination of human aging. It has done so by defining "aging" solely in the privileged "natural" terms of physiological descriptions—even though aging (unlike childhood, for instance, which is terminally marked by the possibility of reproduction at an age that is a relative constant across populations) is a category in which nature and culture are implicated in each other (see Mergler and Schleifer 1983).

In Chapter 3, we inscribed aging within the cultural economy of human life by means of an adaptational narrative response to the question of why human beings live a significant portion of their lives beyond childbearing years. We did so by focusing on the special place that language—and especially narrative language—has in the life of the aged. Instead of "treating the decline in ability with aging as a monolithic entity" and attributing "this decline to either exclusively organic or sociocultural influences" (Spicker 1978: 164), we articulated a conception of old people from the point of view of the complex sender of discourse, both agent and cultural horizon. To do so, we defined "aging" in a semiotic square that described a complicated and complex process that is both physiological and social.

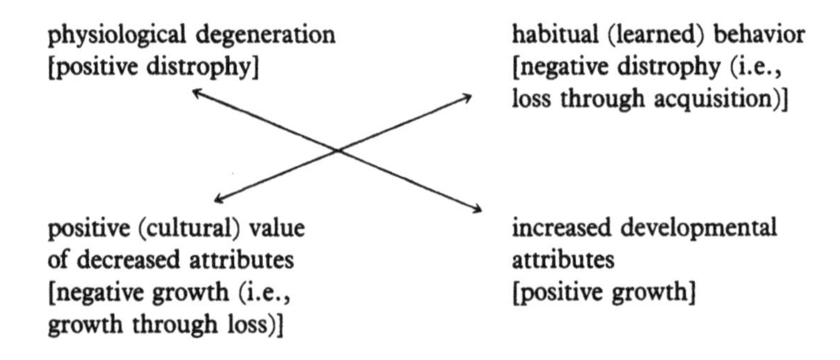

Following these possibilities, as we have seen, old people can be understood as possessing unique cognitive capacities that serve the species. This argument is based on the assumption that the cognition of old people, like enunciated discourse in Derrida's understanding, cannot be understood outside the context of the Other—both other people and the fact, inscribed in the bodies of the elderly themselves, that their age both constitutes and does not constitute their sense of self (Beauvoir 1973; Levinson 1978; Turner 1979).

Similarly, the forms of "strong" and "weak" selection inscribed in this square can be seen as the "explosion" of the cognitive category of "natural selection," whose self-evident simplicity is a basis of the enormous influence of Darwin in cognitive psychology. Weak selection, as may be apparent, functions like the situation of enunciation in which Derrida inscribes language. It suggests a "material" and particular context (as opposed to the transgenerational context of natural selection) which, to quote Lévi-Strauss, remains "latent and submerged, as it were, in time." Such a material context includes aging as the subject of "power" as well as the object of "knowledge." It marks aging as complex. Moreover, it presents a certain logic to aging, a kind of complex, mixed violence in aging similar to the violence that Derrida describes in discourse and that de Beauvoir describes in habit. That this mixed violence of aging manifests itself in the special case of discourse (as well as in the more general category of cognition) is both remarkable and predictable, for the violence of discourse, like so much of the violence of the world, presents, as Yeats said in old age, "bitter furies of complexity, / Those images that yet / Fresh images beget, That dolphin-torn, that gong-tormented sea" (1971: 498).

In Chapter 3 we offered the example of the ninety-year-old man who described the pleasure of talking to his grandson as giving him a reason, at ninety, to get up each morning. That pleasure, as we argued, was that of articulating plotted prose with an explicit moral in order to make experience in time meaningful. In such plotted prose, the violence of aging and the violence of discourse come together. What Donne does for the dying of a good man and Marlow does for Kurtz is to find, in the "story" of dying, a moral to be had, a kind of "thoughtfulness" achieved: thus, goodness makes no noise, and evil can seemingly momentarily transcend itself in the cognitive act of summing up. In both poem and novel, it might be that the narrators are bringing sense to breath and that such events never took place: it might well be that the virtuous man is not speaking to his soul at all—that he is simply breathing—and that Kurtz says nothing as he dies but simply is breathing stentoriously (Schleifer 1990: 192–93). In any case, in these literary examples, as in the familial example, death is inscribed in discourse, both the explicit figure of death narrated in these literary death scenes and the implicit death of not getting up the old man describes to his grandson. Moreover, we can also perceive in these narratives what might be called, following Shelley, the "imageless" figure of death "caught," as it were, in the very breath of language.

That is, these discourses "explode" narrative, as Derrida does with language, into the binary opposition *story vs. moral*. This opposition corresponds to the opposition *strong preparedness vs. weak preparedness* we have described. But more important, it relates the extralinguistic violence of aging—the brute fact of physiological degeneration—to Derrida's analysis of the violence of discourse in terms of the binary opposition of linguistic vs. extralinguistic elements of language. If we bring these oppositions together, we can describe the complex category of the cognition of old people in terms of the opposition *(narrated) knowledge vs. (situated) "thoughtfulness."* Knowledge describes the world: it is discourse-as-cognition in the accusative case. Thoughtfulness addresses the world: it is a form of "power," discourse-as-performance in the vocative case. In old people, these aspects of language are more distinguishable because they are more disproportionate: narrating experience so outweighs other kinds of doing for the elder (weak preparedness), and *its* doing (performance) is so fragile compared to the knowledge it imparts, that it becomes what one old man in a conversation with Blythe called "crotchety" (Blythe 1979: 164). Yet from another vantage, these aspects of language are also indistinguishable for the elder because discourse-as-cognition becomes the last resource of activity (performance); hence, as Wallace Stevens says, just to be in the world would be a way of "saying" something, and saying something is a way of finding reasons to get up in the morning.

Narrative with a moral manifests the complexity of discourse, its cognition and performance. Narrative describes the world, and its moral imposes order upon it. In the Greimassian actantial analysis of genre, the receiver who receives and articulates the moral of the story is always, complexly, another actant within the story itself. In this complexity we can see that the moral of the tale, especially when it explicitly sums up the narrative action with candor and conviction, is the violent imposition of the vocative case upon the accusative. In talking about habit in the aged, de Beauvoir describes this aspect of discourse in terms of its crystallizing power. "When habit is thoroughly integrated into a man's life, it makes it richer, for habit is a kind of poetry. . . . Habit brings about a crystallization like that which Stendhal describes when he is speaking of love: some given object, possession or activity acquires the power of revealing the whole world to us" (1973: 696–97). Such objects "reveal" the world by asserting, in the performance of discourse, a kind of subjective power. Thus, Marlow says of Kurtz, "You should have heard him say, 'My ivory.' . . . 'My Intended, my ivory, my station, my river, my——' everything belonged to him. It made me hold my breath in expectation of hearing the wilderness burst into a prodigious peal of laughter" (Conrad 1971: 49). Yet objects also present cognitive knowledge about the world in the discourse of the elder so that while the discourse of old people imposes the vocative upon the accusative in its very performance, it also inscribes the accusative within the vocative in the knowledge it imparts. For de Beauvoir the repetitions of habit create a language of symbolism that seemingly protects the aged from the anxiety of time and change, but that language also "confers a certain quality upon the world and a certain charm to the passage of time" (1973: 697).

This quality is what Woodward describes in her study of aged poets as "wisdom," and as such it is real, but it is based not so much on the "acceptance" Woodward speaks of as on the required necessity of cognition (or what we called "thoughtfulness") as an activity. This is because the breathing language of the elder is marked by violence: the "thoughtfulness" of age is a function of the insistent context of the violence of aging's decay. The consciousness in the elder of "fierce energy," of "the natural intensity of life," Sally Gadow has written,

> becomes possible through frailty. "It may be a degree of consciousness which lies outside activity, and which when young we were too busy to experience" [Scott-Maxwell 1979: 33]. When physical strength is sufficient for one's aims, when vitality is fully actualized without remainder, there is no conscious access to that intensity. . . . But only when these forms are absent can the pure intensity, the life-force per se, appear in all its strength. (1983: 146)

Here Gadow is describing the relationship between physiological degeneration and the cultural value of decreased attributes, what she describes, quoting Nietzsche, as the "sweetening and spiritualization which is almost inseparably connected with an extreme poverty of blood and muscle" (1983: 146). It is the strange and difficult relationship between Freud's "death instinct" and signification that Coward and Ellis describe, what Lacan calls "the desperate affirmation of life that is the purest form in which we recognize the death instinct" (1977b: 104).

This affirmation manifests itself in discourse: like negation, affirmation itself is a function of discourse. Even in Nietzsche, the figure for this affirmation is a figure for narrative, the eternal recurrence of the same. Old people tell stories, and their stories, like Nietzsche's theory, literally sound verbs more clearly than those of younger people (Mergler et al.: 1985). They do so because, in their *situation* of growing old rather than having their breath latent and submerged in time, time itself is made manifest in their breath: their bodies "speak" in ways that Donne couldn't have meant. The violence of aging leads to speech; it can lead nowhere else precisely because the "consciousness" Gadow speaks of manifests itself in discourse-as-cognition when old people can do little else. For this reason, the complexity of language is made "plain," as Stevens said, in the discourse of old people: "After the leaves have fallen, we return / To a plain sense of things" (1954: 502). According to Stevens, "It is difficult even to choose the adjective / For this blank cold, this sadness without cause," yet his poem, "The Plain Sense of Things" chooses its verbs to narrate, in a plotted text with a moral, the story of the imagination, whose absence, he writes, "had / Itself to be imagined." Nietzsche's figure of eternal recurrence also manifests such "plainness," and in this respect it is very close to de Beauvoir's figure of habit. Both serve to join an acceptance of life—what Nietzsche calls "this ultimate, most joyous, most wantonly extravagant Yes to life" (cited in Gadow 1983: 145)—and the will to power. That is, Nietzsche's work, postulating eternal recurrence yet articulated in aphorisms—which are the formal presentation of explicit "morals"—"explodes" discourse in Greimas's sense of the word. It situates its own cognition by presenting "ideas" nonsystematically, as occasioned and separate rather than a logically coherent apprehension or cognition of the true.

Accusation and Vocation: Two Narrative Rhetorics

Such *situating* of discourse as the complex articulation of both story and moral (performance and cognition) is plain in the discourse of old people. One old man of seventy-nine (whose claim of being happy in old age is the most pronounced of all the elders interviewed in *The View in Winter*) told Blythe the story of his father's chair in which the implied moral is crystallized in an object of the world.

> Well, Father'd set beside me evenins-like and he'd whittle away at things. . . . It was a pleasure to see it. So there he'd set, in his ol' chair—Father's chair, we called it. That wouldn't dew to let him ketch you with your

> arse in it, that wouldn't! I would love to know where that chair is this minute, that I would! . . . Silly fule, I give it away years agoo. The chair Father made. I see him makin' it, an' I give it away! Pity. Father's chair—fancy me a-thinkin' o' that now! But that's how it is when you're an old un, it all kind-a starts up agin, the long agoo. As plain as lookin' out that winder. So this ol' chair. . . . That was Father's [work]bench, that chair. Snares, he'd make. I lay there on the couch larnin' the carvin' and the snares, a-pickin' it all up, gittin' like Father, gittin' Father's skill. (1979: 46)

What makes this man's old age "the happiest time of my life" (1979: 45) is, as he says later, working at his bench as his father did. Yet "working" there is really remembering: "I set at this bench and I don't want the day to end. And it all comes back to me. . . . I'm a boy and I'm by that river, an' we're all there like we used to be" (1979: 49). His father's chair sums up his life, and it also implies the moral of not giving anything away. In the end this old man says of the things he makes, "at fust I'd make 'em and give 'em away, but now I keep everything. . . . I don't copy anything, I make what I remember" (1979: 51).

His father's chair, then, is both experience and meaning. It is almost the crystallization of meaning in that the moral it suggests—a moral that is somewhat small-minded and mean—is that experience and memory are valuable in the most literal economic sense, and that they shouldn't be given away. The lost chair of the father figures this moral in its very absence, just as his narrative articulates it in its presence. Such is the plotted prose with an explicit moral we are describing as the discourse of old people. To figure and imagine it in relation to an object in the world is seemingly to favor the accusative over the vocative and emphasize violence of cognition—the imposition of meaning—over worldly violence. It asserts knowledge—interpretative violence—against the world, as Kurtz seemingly does in his summing up of his experience in a kind of Yeatsian rage.

For Yeats in old age, such knowledge seemed all that was left.

> Speech after long silence; it is right,
> All other lovers being estranged or dead,
> Unfriendly lamplight hid under its shade,
> The curtains drawn upon unfriendly night,
> That we descant and yet again descant
> Upon the supreme theme of Art and Song:
> Bodily decrepitude is wisdom; young
> We loved each other and were ignorant.
> (1971: 523)

In Yeats's aged discourse "some given object, possession or activity," as de Beauvoir says, can reveal the world. Sometimes it is a stone or sword, sometimes a story, often a swan: it is an image that becomes his discourse's moral, as his father's chair serves Blythe's old storyteller. For others cited in Blythe, it can be (as it sometimes seems for Yeats) their very youth. One respondent told Blythe that his youth "was an incredibly long time ago and I look back on an entirely different person" (1979: 187). Another narrates the birth of a young boy and the subsequent death of his mother many years ago, only to add in passing: "this baby boy who was me was called Richard Richardson Pipe" (1979: 62). And finally another respondent tells Blythe, "there is one thing about being very old, you can see what happened" (1979: 227).

In "After Long Silence" such knowledge inheres in both discourse itself after long silence and the lighting of the room. Both light and night are unfriendly: light reveals age inscribed in the faces of these old lovers, which implies that the theme of Art and Song is, as Stevens said in age, "poverty's speech"; while night figures that decrepitude and death which, like that which faced Scheherazade, narrative discourse wards off. In this narrative, discourse is marked by those "Others" who are addressed: the accusative Other of self and old lover, meaningfully marked in their otherness on their very faces and described as decrepitude; and the invoked Other, that extralinguistic violence—that materiality beyond signification (see Coward and Ellis 1977: 148)—that Yeats's narrative betrays and reveals. Yeats's poem, like his poetry in general, narrates the cognition of aging in the violence of its binary juxtapositions—speech and silence, light and darkness, wisdom and ignorance—which mark, as so many of Yeats's poems of old age do, bodily decrepitude *against* the self.

Such violence characterizes the discourse of Yeats's age—Woodward notes that she specifically leaves Yeats out of her account of aged poets because of this violence (1980: x-xi)—which, like the old man in Blythe and like Nietzsche, fiercely asserts the self and its power of knowing against the world: "The world knows nothing," Yeats wrote two years before he died, "because it has made nothing, we know everything because we have made everything" (1961: 510). Yet his speech in age, more often than not, takes the form of narrative, more specifically, plotted prose with an explicit moral. Thus in the second section of "The Tower," when Yeats sends his imagination forth to call "Images and memories / From ruin or from ancient tree," it returns with narrative discourse instead, the stories of Mrs. French, Raftery, Hanrahan, all stories that function to displace remorse: "if memory recur, the sun's / Under eclipse and the day blotted out." Here indeed the old man is finding a reason to get up each morning in the face of aging. Yeats ends the poem by writing his will and "making" his soul:

> Now shall I make my soul,
> Compelling it to study
> In a learned school
> Till the wreck of body,
> Slow decay of blood,
> Testy delirium
> Or dull decrepitude,
> Or what worse evil come—
> The death of friends, or death
> Of every brilliant eye
> That made a catch in the breath—
> Seem but the clouds of the sky
> When the horizon fades,
> Or a bird's sleepy cry
> Among the deepening shades.
> (1971: 416)

Yeats rhymes "death" and "breath": the worst evil is the destruction of the brilliant eye that links sense and breath and makes (or has made) experience in time meaningful. Yeats attempts to speak aging in the accusative, as a "case" of speech, a kind of oracular "utterance." He attempts to achieve knowledge, the kind of transcendental cognition that we can find in Whitehead, and his rage throughout his late poems is the rage against the seeming accident of age—as accidental as it would be had some boys tied it to him. Yet his "knowledge" and its crystallization in a symbolic discourse are also an *accusation:* a kind of will to power against what is.

The symbolization of Yeats's discourse, like that of the old man at his workbench, offers the more or less explicit violence of accusation—the interpretative violence of the "moral" of the story. However, there is another aspect the discourse of age makes plain that inscribes more quietly, perhaps simply invokes, the violence of the world against the human, where the moral is less explicit, less *asserted,* in a narrative that grows out of the situation of aging. Unlike the "knowledge" that age brings to Yeats, this is a form of "weak" preparedness, something that age invokes and traces by taking away everything else. Shelley offers a fine image of such "weakness" in "Hymn to Intellectual Beauty" in which the ghostly shadow of Intellectual Beauty nourishes thought "Like darkness to a dying flame!"

(1914: 527); and Florida Scott-Maxwell describes it in her late autobiography, *The Measure of My Days,* in describing the "fierce energy" of aging: "It has to be accepted as passionate life, perhaps the life I never lived, never guessed I had it in me to live" (1979: 33).

This, too, can be narrated, as in the answer a woman of ninety gives Blythe when he asks if she thinks about death:

> Well, naturally. You have to when there's a dog, don't you? I tell myself I'm thinking about death, but what I'm really thinking about is the dog. Death—dog, that's the way my mind goes now. I'm so old, you see. Terrifically old. . . . Not that I'm frightened by death, oh no. Dear death, how I look forward to it. I look forward to it because I am so tired. So weary, you know. This tiredness just falls on top of me like a dead weight. Such utter, utter weariness—you have no idea. It's worth talking about because it is quite something. I mean I've known what it is to be really tired, like everybody else, but never a tiredness like this. If I tell you that it is too important a tiredness simply to sleep through, you'll know what I mean. It keeps me wide awake so that I can feel every bit of it. I lie on the sofa—or even get the tea—in this huge, huge tiredness. It's not exactly unpleasant, but it is becoming a nuisance. I wouldn't mind if I could just doze off and wake up and find it gone, but I can't. That's not its little game. Being so old is very funny business. To tell you the truth, I can't make head or tail of it! If I get gaga, take no notice, it will be this tiredness. . . .
>
> Let's be serious. This weariness is death. Don't you realize what death is? It is a lovely mist which takes us away. (1979: 232–33)

This old woman is attempting to articulate or name an "other" that is beyond the resources of language and cognition. Whitehead says that "no statement, except one, can be made" of the "remote occasion" of such otherness, namely the statement of ignorance (1967: 25). But this old woman, like the discourse of old people more generally, attempts an articulation by means of a different mode of cognition: she can speak of it only by speaking to it, by playing games with it, by not sleeping through or ignoring it. As Derrida said, "I can only, I *must* only speak to the other; that is, I must call him in the vocative, which is not a category, a *case* of speech, but, rather the bursting forth, the very raising up of speech" (1978: 103). Here the violence of language is not the interpretative, accusatory violence of cognition, of a story's moral, but a kind of violence "behind" language, invoked as this woman attempts to articulate death by speaking of her dog, her weariness, and the vague mist she mentions. Like that tiredness and that mist, it is not a thing at all but only something that is there all along, behind speech, which darkness would let shine forth. It is the part of living life's business and energy never allowed her to guess she had in her to live. This respondent plots this all but unconscious awareness in a narrative of her day—nighttime, tea time, resting during the day—yet the "moral" that emerges here is not in or as some thing but is invoked, as an optical illusion is invoked, by the arrangement of its elements, the fact that speech is "raised" at all.

This invocation is perhaps clearer in the quiet narrative discourse of Wallace Stevens, whose poetry—seeking as it did in his old age the plain sense of things—reveals a violence less explicit than Yeats's, but one in which, we think, the violence of aging is sharply delineated nevertheless. He articulates what the old woman is attempting, the "vocation" of age, addressing that other where only vocatives—shouting, crying, gesticulating—are possible, that other without system that can only be addressed.

> That would be waving and that would be crying,
> Crying and shouting and meaning farewell,
> Farewell in the eyes and farewell at the centre,
> Just to stand still without moving hand.
>
> In a world without heaven to follow, the stops
> Would be endings, more poignant than partings, profounder,
> And that would be saying farewell, repeating farewell,
> Just to be there and just to behold.
>
> (1954: 27)

The violence of Stevens's poem is the violence of dying embedded in our lives, just as the all but unconscious awareness of materiality—the resistance to signification Lacan calls the "Real"—is embedded in any cognition, no matter how abstract, by means of the material objects that manifest that cognition. Such violence is quieter, without Kurtz's words and rage (did he say them?) or Yeats's rage, yet it inhabits all things by virtue of their more or less acknowledged otherness. Stevens's is a vocative violence—a violence of vocation—as opposed to Yeats's accusative and accusing violence. "To be one's singular self," Stevens goes on, "to despise / The being that yielded so little, acquired / So little, too little to care, to turn / To the ever-jubilant weather . . . That would be bidding farewell." Here is not the acceptance of wholeness but the lamentation of fragmentation, the poignancy that comes from stops, the beauty that comes from decrepitude and death.

Death marks Stevens's discourse: as he says in "Sunday Morning," "death is the mother of beauty." The poem we are quoting, "Waving Adieu, Adieu, Adieu," does not present, as Yeats so often does, a story and a moral but, rather, attempts to create an atmosphere and a mood (an atmosphere or "mist"), to remark in its final question the junction of self and earth we mentioned earlier, just as Yeats's early poetry attempts—with a great deal more violence than Donne's virtuous men—to mark the disjunction of self and earth: "Ever jubilant," Stevens ask, "What is there here but weather, what spirit / Have I except it comes from the sun?"

In old age Stevens came to see that questions are remarks, and like Yeats, he transformed a poetry of moods to one of narrative. "Questions are Remarks," for instance, offers the elements of a complete narration so that when his grandson, age two, asks "What is the sun?"

> His question is complete because it contains
> His utmost statement. It is his own array,
> His own pageant and procession and display,
>
> As far as nothingness permits . . . Hear him.
> He does not say, "Mother, my mother, who are you,"
> The way the drowsy, infant, old men do.
>
> (1954: 462)

The discourse of age marks what "nothingness permits" by invoking the terror of death behind life while it narrates and describes the fullness of the life of his grandson, "the expert aetat. 2": "In the weed of summer comes this green spout why." It does so, as Yeats did, by superimposing the accusative and the vocative—by performing the violence of including plotted prose within the vocation of explicit moralizing (the vocation of cognition)—in the discourse of the elder. Yet Stevens's discourse is different from Yeats's. For him, stops are profound endings, Derrida's "unimaginable night which would not be the opposite of nonviolence: nothingness or pure non-sense" (1978: 130), against which the violence of accusation seems small. Whereas Yeats's symbolism is the "knowledge" that age brings, Stevens's "knowledge" was there all along. It came with the very raising of speech, seen now because, for the aged, speech is all that is left.

That is, Stevens *narrates* death in the "stops" of moralizing—the stops of the interpretative violence of cognition—by invoking its continued, speechless (infant) presence: its inaccessible presence as Other. The questions of old men invoke and remark the nothingness inscribed in speech, and Stevens narrates that violence by telling the story of the strength and limitation of his grandson: the child says, "Mother, what is that" and while the mother explains—the sun is pulled across the sky by red horses, she says—the old man's drowsy speech both answers and does not answer: the infant speech of children and old men come to the same thing, neither speech nor silence, a remarking, not remaking, of the world.

It is a world, different from Yeats's or Greimas's, in which the sun does and does not rise with rhetoric. It is a world, that is, of ever-jubilant weather which violates us as much as that complex accusatory noise—of speech *and* silence—of the tin cans of old age. Which Yeats proclaimed in old age that the poets made everything, Stevens's lesson is different. The last lesson of the master, Roy Harvey Pearce has argued, is the sense of weather, and he quotes Jorge Guillan to articulate it: "I am, but even more, I am here, I breathe. What is profound is the air. Reality invents me: I am its legend" (1965: 135). Here again breath and sense are linked, and again we can inscribe the discourse of the elder in a semiotic square:

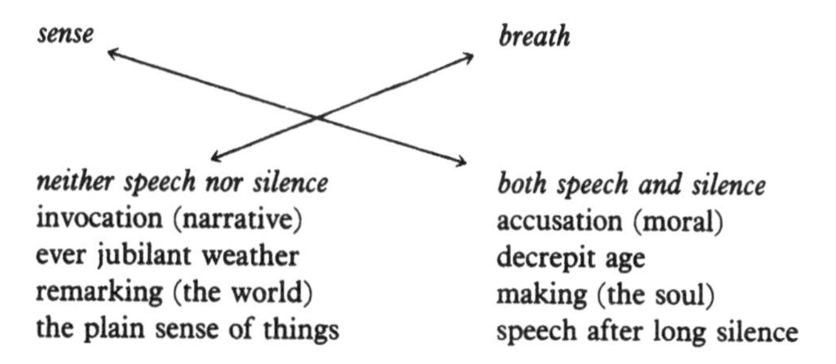

The Discourse of the Aged

The plain sense, then, is the interplay of linguistic and extralinguistic aspects of language—located, like Lacan's "truth," Greimas's narrative, and the "comfortable" accommodations of social cognition, in the opposition of sense and nonsense—that age makes plain in discourse in the two ways we have described. In "To an Old Philosopher in Rome" Stevens narrates an eloquent description of old age—"a kind of total grandeur at the end"—in a combination, as forms of weak and strong adaptation, of contemplation and experience that presents the discourse of old people. The voice in Stevens's poetry combines in strange and haunting ways the authority of experience, the strong sense it gives of objects in the world, "a bed, a chair and moving nuns" and their interpretation, along with the hesitations and modulations—in a word, the "weakness"—of "thoughtfulness." The situation of thoughtfulness in his poetry is invoked by the constant repetitiveness of language which combines a need, almost an anxiety, to speak, and the awareness that there is little else it can do. With this voice, its dative vocative mood, Stevens "narrates" old age with its moral:

> And you—it is you that speak it, without speech,
> The loftiest syllables among loftiest things,
> The one invulnerable man among
> Crude captains, the naked majesty, if you like,
> Of bird-nest arches and of rain-stained vaults. . . .
>
> It is a kind of total grandeur at the end,
> With every visible thing enlarged and yet
> No more than a bed, a chair and moving nuns . . .
>
> Total grandeur of a total edifice,
> Chosen by an inquisitor of structures
> For himself. He stops upon this threshold,
> As if the design of all his words takes form
> And frame from thinking and is realized.
>
> (1954: 510)

Stevens's images of aging perception in "To an Old Philosopher in Rome" are images, as he says, of truths as they are seen: "the figures in the street / Become the figures of heaven"; "How easily the blown banners change to wings"; "Things dark on the horizons of perception, / Become accompaniments of fortune." These are images, as suggested by the empirical psychological studies we cited earlier, of a different cognitive order than that of adults in possession of their sensory functions. Whether the imagination of age, as Stevens says at one point in this poem, is "poverty's speech" to compensate for lost powers, or whether it is, as he says at another point, a different way of seeing that increases perception into a kind of Yeatsian grandeur at the end, is really irrelevant. In either case, age becomes a locus of violence, weakly or strongly described. It is the discourse of those who depend on cognition; and for whom the accusative and vocative violences of discourse and cognition, like the violences of life, are real.

"There used to be no house," Walter Benjamin wrote in an essay that attempted to define storytelling,

> hardly a room, in which someone had not once died. . . . Today people live in rooms that have never been touched by death. . . . It is, however, characteristic that not only a man's knowledge or wisdom, but above all his real

> life—and this is the stuff stories are made of—first assumes transmissible form at the moment of his death. . . . Suddenly in his expressions and looks the unforgettable emerges and imparts to everything that concerned him that authority which even the poorest wretch in dying possesses for the living around him. This authority is the very source of the story. (1969: 74)

This authority may very well be the source of the cognition of narratives as well, the ability of story to make experience in time meaningful. For Benjamin, the unforgettable emerges from the faces of the dying—from the sense that dying gives of the materiality of our lives: Stevens figures such materiality as weather, just as the old woman Blythe quotes figures it as a mist. Yeats figures it as the knowledge and wisdom born of bodily decrepitude, like the workless time Blythe's old man sits at the bench. Yet from that material decrepitude of old age comes narrative discourse, a telling which simultaneously has its say and waves goodbye—a telling which marks and situates its sender. It is the activity of cognition that serves the species, as we argued in an earlier chapter, and it also serves the subject of discourse and cognition in the ways we have described here. In the next chapter we look at the "transmissibility" of cognitive knowledge more closely—in the institutions of critical reading, pedagogy, and publication—to examine the public and social forms of cultural cognition that we have examined in these chapters by focusing on subjective knowledge and personal narratives.

Part III

CULTURAL DISCOURSE

7

THE INSTITUTION OF CRITICISM

Pedagogy, Publishing, and Oppositional Criticism

The Ethics of Cognition

In this chapter we examine the institutions of cognition —ways in which understandings come to take shape and to be taken to be objective truth—by examining particular modes of understanding that have come to inhabit literary studies and literary criticism in the last two decades. In pursuing this example of the institution of understanding in a particular discipline, we also try to suggest that this institutionalization of knowledge is not confined to a single discipline. To this end, we examine the ideological nature of cognition in Aristotle's "oppositional square"—a model Greimas borrowed, but one which is closer to the Anglo-American empiricist tradition than to semiotic rationalism. In addition, we consider theories and critiques of pedagogy as they have developed in recent years. These theories and critiques have affected literary study but draw on teaching well beyond its scope. Finally, we examine the production of scholarship in the definition and generation of literary knowledge, but we do so in the context of a wider scope of the relationship between ethics and knowledge, the ethics of publishing.

Implicit in this long chapter—and in the asymmetrical relationship of the single chapter of Part III of *Culture and Cognition* to the other parts—is a sense of the complexity of cultural discourses and the cognition they condition. This is perhaps most clear in the *problem* of ethics. Like the complexity of narrative cognition we describe in conclusion to this chapter, the problem of ethics is precisely the problem of the relationship between general and special cases—between moral "law" and particular human actions. Stanley Fish has recently described the complicated relationship between general law and particular action in discussing the laws governing "free speech." Free speech, he argues, never really existed precisely because it is impossible to abstract the cognitive aims of speech from the local interpersonal activities. "Expression," he writes, "is more than a matter of proffering and receiving propositions" since "words do work in the world of a kind that cannot be confined to a purely cognitive realm of 'mere' ideas" (1992: 241–42). Such timeless conceptions as "free speech" and the absolute ethical injunction that speech should never be regulated are thus problematic. "The question of whether or not to regulate [speech]," Fish says, "will always be a local one and . . . we can not rely on abstractions that are either empty of content or filled with the content of some partisan agenda to generate a 'principled' answer" (1992: 243). Even if we replace "law" with "order," as Evelyn Fox Keller does, we are still faced with the problematics of the framework defining "order": as Wallace Stevens says in "Connoisseur of Chaos," "A. A violent order is disorder; and / B. A great disorder in an order" (1954: 215).

To put this differently, ethics is not a detachable "part" of philosophy the way, we have suggested, empirical cognitive science, semiotics, and psychoanalysis can be (and have been) detached from one another in examining the nature of cognition so that our task is to superimpose these traditions on each other. Neither is it detachable from "objectivity" as advocates of "value-free" scientific inquiry often assert. Rather, ethics is itself the activity of superimposing general and special cases, the constative and the performative, knowledge and power, the various "overdeterminations" of psychoanalysis. From the vantage of ethics, the relationship between order and disorder Stevens describes is a recurrent problem, constantly emerging, that no general theory, no disinterested standpoint, no piece of "knowledge," can resolve once and for all.

Still, simply because it cannot be accomplished once and for all does not mean that ethics—or the truths of cognition, for that matter—cannot be the goal of human activity. It does not mean that science shouldn't strive for value-free inquiry. Rather, the work of ethics is the work of culture in its constant aim to develop provisional accommodations governed by the never fulfilled hope for general laws, orders of justice, the coincidence of truth and meaning. The tools of this work are the "explosions" of semiotics that attempt to situate and analyze self-evident truth, the refusal to submerge too quickly parts into "simple" wholes (such as the reduction of old people to a version of adult "humanity" or of the situations of teaching to the "background" of its constative "knowledge"), and the multiplications of narratives that attempt, as Ricoeur says, to explain differences

(1984: 124–25). Edward Said describes such work as both unprovincial and interested under the term "worldliness." Discussing a novella by the Palestinian novelist Ghassan Kanafani, he argues that "worldliness" restores the connection between such local works of art "to a whole world of other literatures and formal articulations." "It seems to me absolutely essential," he continues, "that we engage with cultural works in this unprovincial, interested manner while maintaining a strong sense of the contest for forms and values which any decent cultural work embodies, realizes, and contains" (1992: 185–86). Ricoeur articulates the problem of being both unprovincial and interested at the same time when he asserts that the goal of explaining difference in narrative is part of his larger thesis that "an event, as what contributes to the progression of a plot, shares with this plot the property of being both singular and typical at the same time" (1984: 251). This, we believe, is also the problem of the relationships of cognition and culture.

The three parts of this chapter—on oppositional criticism, pedagogy, and publishing—traverse these areas of concern in an attempt to enact a complicated ethics of cognition, the complexity of cultural discourse. But in the same way that ethics is not "detachable" from philosophy, so its cultural discourse is neither the whole nor a part of the cognitive and performative elements it surveys and articulates. Rather, ethics takes the forms of the explosion of self-evidence, the thickening of description, and the multiplication of narratives in its scrupulous attempt to connect the general with the particular. For this reason, it can make any elements it surveys the occasion for analysis, description, and narrative that can alternatively complicate general truths with special cases or understand and judge special cases against general propositions. Such judgments are alternatively unprovincial and interested, just as semiotics, empiricism, and narrative itself are alternatively cognitive conclusions and occasions for cognitive activity. For this reason, under the rubric of "Cultural Discourse" we offer a generalizing special case of the institution of cognition.

Knowledge and Ideology: Oppositional Criticism

Since the 1970s several American and Continental critics have come to epitomize the possibilities for an "oppositional" or radical critique of contemporary culture, from practices in the academy and arts to international politics. Most prominent are Michel Foucault, Noam Chomsky, Edward Said, and some of the "new" Marxists, particularly Jean Baudrillard (in his earlier work), Fredric Jameson, and Louis Althusser. To some degree, most of the figures in this movement have used the kind of cognitive semiotics we have examined in this book for probing the political dimension of culture. In other words, these oppositional critics respond to cultural experience with a model of cognition based on the dynamic articulation of ideology within cultural discourse.

Such a cognitive-semiotic model, for instance, is consistent with Antonio Gramsci's understanding of hegemony and its dialectical relationship with other modes of cognition. For Gramsci, the generation of political strategies and ideology functions much as cognition does by continually generating logically opposed possibilities, new candidates for the "hegemony" (or dominant ideology) that arises out of social and economic conflicts. From this vantage point, as Gramsci argues in "Art and the Struggle for a New Civilization," the critic who can "represent the given moment" can also represent the "prevailing" historical discourse in the largest sense (1985: 94). Gramsci, like Greimas, argues not that the attempt to "represent" ideology guarantees transcendental understanding but only that the dynamic nature of this process opens the space for critical representation of the social or cultural dialectic. From Gramsci's perspective, the dynamics of cognition we have been examining throughout this book are congruent with the dynamics of ideological and cultural critique. It is a short step, for instance, from Gramsci to Fredric Jameson's use of Greimas's semiotic square in *The Political Unconscious* (1981) as a key to a radical literary and cultural criticism.

In this way, the political use of cognitive semiotics has helped to reshape and redefine cultural criticism. But many critics, such as Barbara Foley and Catherine Gallagher, have questioned the particular blending of politics and semiotics that has taken place. They have asked, as Gallagher does, whether oppositional critics can properly "derive politics from the nature of criticism itself" (1985: 41; see Foley 1985), suggesting that ideology does not inhabit the very instruments by which we know the world. Gallagher and several other critics of oppositional criticism deny that ideology (politics) exists meaningfully at the level of "critical" understanding, certainly not at the level of semiotics and cognition. If they are right, if there is no intrinsic relationship between politics and cognition, then the "engaged" critic (like Jameson or Said) is merely inserting ideology where it does not belong, merely importing politics as a rhetorical overlay to criticism and possibly, in the process, obscuring the "real" questions of political struggle in the world.

The oppositional critics assert, however, that politics and criticism presuppose each other. While critics such as Said, Chomsky, and Terry Eagleton tend to affirm the critic's ability to know and represent historical challenges to hegemony—the ability of the critic to occupy an "outside" space from which a hierarchy may be opposed and disrupted—others, like Baudrillard, see the critical activity (and the cognitive activity of cultural critique) as complicit with ideology. Baudrillard is more concerned with the indeterminacy of intervention, with limited politics and the traps of ideological error. He even criticizes Marxism for failing to escape, to generate oppositions beyond, its pervasive "labor/production model" (1975: 29). That is, many Marxist critics (particularly Jameson and Baudrillard) now argue that the politics of the new cultural criticism, based largely on cognitive semiotics, is not a cosmetics of "engaged" rhetoric but precisely ideology inscribed within understanding—"politics" that, in fact, can be derived from the cognitive nature of criticism itself, the networking and underwriting of understanding we examined earlier. In this argument, they suggest that ideology, manifested in formal notions of "opposition" and "contrariety," is never absent from language and is a political force at the deepest levels of semiosis, ideology being, as Baudrillard says, "the very form that traverses both the production of signs and material production" (1981: 144). At stake here, as Gallagher, Said, Foucault, Chomsky, Paul Bové, and many concerned with oppositional criticism from different sides acknowledge, is the horizon of cultural politics seen within the institutions of literary studies and the academy.

Here we are focusing on the theoretical implications of the seemingly formal oppositions of cognition—the binarities of understanding with which we began our study and which we have examined in different situations of cognition—in order to argue that, as the new oppositional critics assert, "ideology lies . . . in the internal logic of the sign" (Baudrillard 1981: 144). Cultural critique, it follows, must necessarily be a semiotic/political reading of cultural practices—much like the narrative readings of cognitive practices examined in earlier chapters. Underlying oppositional criticism is the set of assumptions about what conflict is, why some things are cognitively "opposable" and others are not, assumptions we have traced in schemas of cognition throughout this book (most often under the rubric of "narrative").

The tradition of thinking about opposition, from the pre-Socratics and Aristotle through Hegel, Marx, Gramsci, and Baudrillard, tends to posit knowable rules that are thought to govern the possibilities for conflict and opposition in a given historical instance of cognition. In this thinking, we can see the strong structuralist and semiotic leanings in such theory, and we can see that the *forms* of cognition we have studied also encompass forms of political consciousness and, as in Jameson, political unconsciousness as well. This broad approach from Aristotle through Jameson frequently advances a four-stage analysis, closely related to Greimas's semiotic square, that is intended, in theory, to encompass and exhaust possibilities for the relations of conflict. A strong example of this political (as related to cognitive) formalism is Jameson's political use of Greimas's early social definition of the semiotic square. In the "Interaction of Semiotic Constraints" Greimas describes the analysis of the semiotic square as always a social analysis: "it is a question," he writes, "in the case of nature as in that of culture, of social values (and not of the casting of nature into the realm of nonsignification)" (1987: 54). That is, Greimas sets up the square to analyze any social fact within a four-term homology to the end of exposing its "problematic," or the ideological framework that creates it, the alternatives it presupposes, in a network of understanding.

We have already examined the square in terms of the reasonableness of cognition in the Introduction (as well as in a host of other situations). There we used the term "intuition" to describe the contradictory to "reason," even though such a term is complexly figurative as well as abstract. This opposition between abstract designations and semantically charged designations in the semiotic square creates a constant tension in Greimassian analysis. We saw such tension in Chapter 2 when we "filled in" the abstract designations of the square with actants and in the preceding chapter when we abstracted forms of understanding—rhetorical strategies—from empirical data concerning aging. This difference between logical categories of thought and semantic categories marks the *ideological* functioning of the square and, more important, of cognition itself. It is for this reason that in the "Interaction of Semiotic Constraints" Greimas offers both a "logical" model of the square and its "semantic" investment (1987: 54; see Schleifer 1987a: xix–xxiv for a fuller discussion of this opposition).

In fact, the square's second level constitutes a formal recognition of the ideological dimension of cognition. The second level, as in our example in the Introduction, can place (or "configure") the abstract, logical relationship of the first level into history. If the logical contrary to /reason/ is /irrationality/, the opposition between /discourse/ and /intuition/ on the second level *situates* that logic within its historical manifestations. As we saw in Chapter 2, the square superimposes constative and performative language, *recognition* underlying *inquiry* (the "inquest"), *marking* underlying *information*. Jameson himself notes, despite the seeming formalism of the semiotic square,

> in actual practice, however, it frequently turns out that we are able to articulate a given concept into only three of the four available positions; the final one [the fourth position], remains a cipher or an enigma for the mind. . . . At this point, therefore, the development of the model may take two different directions. It may involve the replacement of the abstract terminology with a concrete content. . . . Or it may take the form of a search for the missing term . . . , which we may now identify as none other than the "negation of a negation" familiar from dialectical philosophy. It is, indeed, because the negation of a negation is such a decisive leap, such a production or generation of new meaning, that we so frequently come upon a system in the incomplete state . . . (only three terms out of four given). (1972: 166)

For Jameson, then, there is no cognition whatsoever, and no semiosis, outside the socially derived four-term homology. In other words, the fourth term (/discourse/ or /marking/ in our early examples) is always situated socially, and, so situated, it marks and discursivizes the ideological nature of cognitive activity.

For this reason, Jameson sees the semiotic square as always constituting the crucial "nucleus" of a whole "ideological system" (1981: 254). It is dialectical in operation and exists under three semantic horizons, those of "political history," "social history," and panoramic "history" (or History as such) (1981: 75). The square is a "nucleus," as Jameson explains, in that by mapping "the limits of a specific ideological consciousness" it is the "graphic embodiment of ideological closure as such" (1981: 47–48). A literary text positioned on the square, thus, will advance the homology of a "concrete social contradiction" (1981: 254) and, in this way, "tilts powerfully . . . into this very political unconscious" of potential ideological positions (1981: 49). By this, Jameson means that as the second level of the square explicitly formulates ideology for the first, ideology on the square can then be seen in relation to other signifying systems. Jameson, in sum, sees the square (and semiotics itself) as a "symptomatic projection . . . of social contradiction" (1981: 83) that can then be understood in "historical" terms—that is, in a dialectical relationship with other semiotic/political systems. In this way, the square marks the ideology of cognition.

If we use Jameson as a model, it is evident that one task of the oppositional critic is to exhaust the logic of narrative discourse within an ideological frame (a "problematic") and then to submit the ideology to a historical reading. Jameson does this by seeking resolutions for "concrete social contradictions" that the square is uniquely capable of making evident in its "foregrounding." In this way, he could put Marxist dialectical thought in modern industrial society on the square in relation to a "bourgeois" understanding. The bourgeoisie, representing the dominant social practice, would then oppose the proletariat (in a "contrary" relationship) on the first level. On the second, the petty bourgeoisie (in a third position) would oppose the revolutionary vanguard (the fourth position):

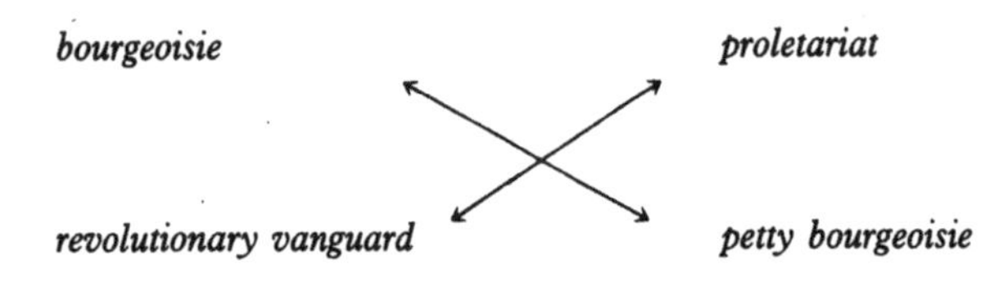

This analysis theoretically "exhausts" bourgeois logic by exposing the suppression of revolution (represented in the fourth position of the square) as a condition, a cognitive requirement (or "rule"), of bourgeois dominance. It is this suppression that makes thinking the fourth position, as Jameson says, so difficult. On the square, the ideology of bourgeois society is marked by the third position, the fact that the *values* of the petty bourgeoisie (which, in Marx's analysis, is *both* bourgeois in its thinking *and* proletariat in its socioeconomic position) contradict the socioeconomic position of the petty bourgeoisie. The bourgeoisie, then, as Marx and Engels argued forcefully, both generates and suppresses revolution. In this prototype of an oppositional criticism, ideology is the dynamic aspect of analysis, simultaneously emerging from and then demonstrating the potential to disrupt a cognitive practice.

Other oppositional critics roughly fit the Jameson model. Derrida (certainly for some an "oppositional" critic) and Foucault disturb signifying practices in Western epistemology by foregrounding "structurality" (Derrida) and "discourse" (Foucault) as virtual "fourth terms" for the dominant terms "presence" and "knowledge." Edward Said, as both a critic and a Palestinian Arab in exile, may be, as Catherine Gallagher points out, the best model of the oppositional critic in that he represents, as she says (even while disapproving of Said's critical strategy), "the paradigmatic attempt at integrating the roles of literary critic and political advocate, at giving them a joint foundation" (1985: 37). She means that the "fact" of Said, at least as he presents himself, is a representation of ideology within American critical practice in that by belonging to an "outside" cultural and political discourse his presence creates the space of "exile," the opposing position that defines the dominant practice. Gallagher rejects this conflation of politics and criticism, arguing that political choices cannot be found in the reductively "analytical" space of criticism (1985: 42). In this manner, she rejects the ideological "content" of cognition.

Said's position, however, as American-Palestinian and leftist critic, would necessarily generate a second semiotic level in American criticism that exposes the relationship of history and criticism in the dominant practice of American culture in such books as *Orientalism* and *The World, the Text, and the Critic*. The point is not so much that anyone should derive politics from literary criticism, but that in the interaction of the two across the levels of the semiotic square, ideology is shown to be, in fact, inherent in the cognitive activity of criticism, as it is in all semiotics. Still, the question remains—and this is perhaps Gallagher's and Foley's point about such critics as Said—concerning the political efficacy of this syncretized position. The idea of "politics" inherent in the sign and in the oppositional stance does not constitute a political program or an effective difference. Foucault, similarly, attempts to identify the limits of personal intervention so as not to be led—as he thinks Chomsky is—into ideological error in immediate political gestures too bound up in the ideological residue of unintended stances and aims to effect goals worth achieving (see the interview with Chomsky and Foucault in Elders 1974). Whereas an oppositional critic like Said attempts to "expose" ideology, to chart and enhance disruption within a socially sanctioned discourse, Foucault imagines the overdetermination and epochal operation of discourse to extend largely beyond the reach of personal intervention (Elders 1974: 174). The nature of discourse, therefore, necessitates a kind of "micro-physics of power" (Foucault 1977a: 139) and implicitly argues against a grander politics, or what Lenin called "revolutionism"—enthusiasms and fantasies about intervention regardless of their actual effectiveness. From Foucault's viewpoint, also, there is no simple disruption from "outside" a discourse because there is no simple "outside," since all movement in discourse is constituted by discourse and its disruptions. Said, it follows from this point of view, cannot be in cultural exile as a critic because no such space as "exile" exists, except as positioned ideologically within the dominant discourse.

Foucault, of course, grants the ideological dimension of the sign and of cognition that we have been discussing, but precisely because of the mutual engagement of cognition and politics, he cautions against believing that one stands clear of ideology enough to "oppose" it directly, as Said and Chomsky seem to think is possible. Foucault's argument—on its face suggestive of political quietism—is that the indeterminacy of ideology, in any realm except the most immediate microphysics of power, precludes political intervention of the kind Said and Chomsky advocate.

A closer look at early historical articulations of the formal mechanisms of "opposition" we have been tracing reveals that there is good reason for critics to differ over the social effectiveness of a cognitive-semiotic oppositional criticism. Aristotle, like oppositional critics, posits a common foundation for language, cognition, and politics (especially in relation to rhetoric and propositional logic), and he is still a relatively thorough explorer—in "On Interpretation," "Categories," and *Posterior Analytics*—of the complex of cognitive "conflicts" inherent in opposition. Aristotle's "oppositional square," in fact, is the prototype of Greimas's "semiotic square" and of Lacan's Schema L. He developed the square to account for change and show that cognitive possibilities must move along "oppositional" lines or else change—and especially political change—would be beyond understanding. As John Peter Anton says, without such oppositional lines changes "cannot be accounted for" because they will be lost in the endless expanse of heterogeneity and chance (1957). For this reason, in the *Posterior Analytics* Aristotle asserts that "there can be no knowledge of that which [exists or comes to be] by chance" (1981: 42). In Aristotle, as Anton notes, contrariety as a logical formation serves to render difference intelligible (1957: 86). Aristotle's analysis of opposition argues for, much as Baudrillard's and Greimas's analyses explicitly show, the reciprocal presupposition of ideology and cognition and, further, the susceptibility of both to difference and the infinite regress that difference opens up.

The semiotic approach to oppositional criticism, in short, precisely threatens substantialist notions of ideology (and the tradition of such criticism within the political left) by making ideology as subject to difference as any other discourse. This "difference," as Aristotle's "On Interpretation" shows, shifts the ground for the conception of values and ideology, and so Gallagher and Foley are correct to isolate this point as the distinctive feature of the new political (literary and cultural) criticism.

Aristotle marks this shift by using a *political* example in the demonstration of the oppositional square of cognition in "On Interpretation" (1962: 48):

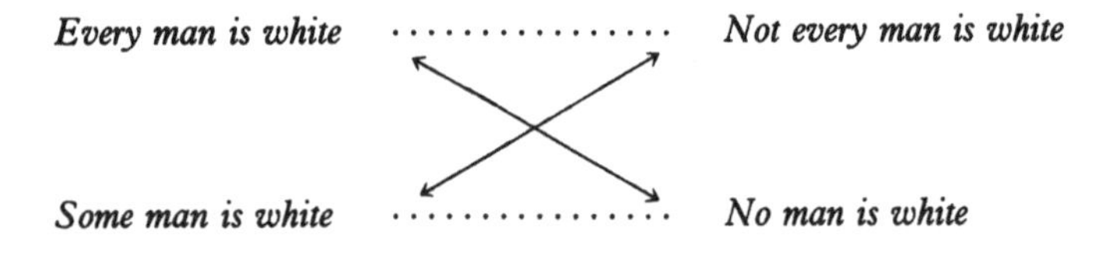

On the first level, the square begins with the proposition that Aristotle calls *hupokeimenon,* which means "subject" but also "substance" (*ousia*). This "subject" is the substratum or underlying form of an argument that impels the square dynamically as "subject-in-progress" (Anton 1957: 11)—"subject" being both the angle on meaning and "meaning" as articulated in the square's positions and relations. The square as a whole represents the "positions" and "relationships" of cognitive meaning, but it begins with the posited ("given") speaking subject, or subjectivity, established by the first position (here the one that says "Every man is white"). That is, it begins with self-evident "truth." The first proposition, then, while not yet an articulated, "relational" term, necessarily embodies the potential for elaboration that will unfold as the four corners of the square.

The second position ("Not every man is white") then establishes, as Anton explains, "limits and extremities" for the square's subsequent relations (1957: 57). The second term creates a "set" encompassing the initial proposition. In this case, the second term establishes the set composed of white men and nonwhite men, but it banishes the exclusive possibility of no differences, *only* white men. In this way, the second term creates a significative context, a range of difference larger than and not limited by the first term. With the introduction of the second term, the set without differences (what Aristotle calls *isonomoi*—"citizens" or "men who are the same") is destroyed, and the original term ("white men") then reemerges within a newly articulated range of possibility and difference. The "contrary" opposition of the first level, therefore, succeeds in marking off a limited set (or range of reference) against the background of unlimited difference. This range is what we called, in Greimas's term, the "axis" on which the two terms of the first level exist, now seen as the locus of a culturally determined semantic network.

The third term (on the second level) limits the range of reference even further. This proposition "contradicts" "Every man is white" with the exact opposite, "No man is white." It confronts presence with absence. The cognitive possibility of men who are not white is generated by the contrariety of the second term, and now the square narrows exclusively to that option to form the contradictory negation or reversal of the first term. With the third term it becomes clear that the square is narrowing its range of reference, focusing ever more tightly, as it moves from proposition to proposition away from the originating first term.

The fourth term completes this narrowing with "Some man is white," which specifically contradicts the second proposition of "Not every man is white." This farthest reach of the square completes the square's logic, but it also goes beyond the other oppositional relations. That is, the first-level relationships signaled the coexistence (but not simultaneously) of contrary possibilities. The contradictory relationship of the third and first terms then signals the privation or absence of the other. The fourth term, finally, has no oppositional relationship with the first, but as "contradictory" to the second it is a selection made possible by the range of extremes created on the first level—a selection that is not restricted to a negation (as in the third term). Greimas describes this relationship between the fourth and the first terms (and that between the third and second terms) as one of presupposition, and it is this contention that distinguishes his semiotic square from Aristotle's oppositional square. But in both cases, the fourth term thus marks an extreme degree of specificity—of narrowed range—for the square in relation to its first term. It marks a new set of heterogeneous possibilities only distantly tied to the first term.

In fact, the Aristotelian square emphasizes that the fourth term is more remote and freer from the first term than may at first appear. All of the other oppositional relations in the square were dictated by some degree of exclusion. The fourth term, though, as the end of that process, is a selection from the heterogeneous range of possibility created by the oppositional poles of the first three terms. Those poles, in fact, form a kind of scaffolding of opposition gradually rising above, suppressing, and escaping the expanse of heterogeneity they governed. Now, "Some man is white," like the "revolutionary vanguard" in the previous example, comes into the square as the specific instance, the necessary and yet disruptive example of heterogeneity that the square all along was suppressing through its insistence on various formal oppositions. The square's logic, then, generates the fourth term but has all along suppressed the very possibility the fourth term represents—in this case that a white man is a special case (rather than the defining case) of humanity.

What we see in this example is Aristotle's demonstration of how ideology (and the problematics of ethics) is produced. That is, each proposition of the square marks off and narrows a previously larger range of reference, which, in turn, enables a further narrowing. The oppositional system these narrowings produce is, to some degree, merely formal and artificial. The square's oppositional "system," however, as Anton says, has the effect of rendering difference intelligible by submitting "all the differences to . . . the principle of contrariety, which in turn becomes the pivot-point for relating, organizing, and systematizing difference" (1957: 86). The result is a "system" of values that is represented by the square's propositions and that is drawn from and thus escapes the unintelligibility of mere differences or unorganized heterogeneity. The system of value escapes unintelligibility by implying a narrative order we have made explicit in this analysis. For this reason, in the necessary specificity of semantically charged terms, it escapes the logical formalism of abstract syntax. In its narrative form—neither pure contiguity nor logical implication—we can see the intersection of cognition and ideology.

In other words, as the appearance of the fourth term underscores, the square has constructed ideology out of the very range of differences it marked off through its own delimitations. Aristotle's demonstration shows, as Ferdinand de Saussure and Greimas note, that ideology is articulated within a differential system, differences without positive terms, and without a fixed reference. Any ideological formation produced by the square can potentially be resubmitted to its operation, and a new ideological transmutation is possible. Still, as we have seen, the absence of fixed reference does not mean the absence of any reference. The square itself fixes reference by beginning with the self-evident truth of the semantics of the first term. The fourth term, through its specificity, unfixes that reference by reinscribing a critique of hierarchy in the square. It implicitly returns the square to its own suppressed origin—a special case of understanding taken as the defining case. Thus, it "unfixes" reference but does not (necessarily) abolish it. Rather, it—as the square as a whole—makes reference a cognitive activity subject to narration.

What we are viewing here, as much as with Jameson, Greimas, and oppositional critics generally, is the application of semiotic analysis to the cognitive understandings of ideological systems of value. In this view, ideology—and understanding—can no longer be an immutable category of belief but is an "effect" of a system of differences and is continually subject to transformation within that system. For instance, the starting place of Aristotle's exemplary square in "Every man is white" is not ideologically innocent. As Page duBois shows in her analysis of Greek art and ideology, "male" and "white" are revealing selections from a set of oppositions generated by Athenian culture (just as "conjugal love" is an interesting beginning point for Greimas in his articulation of the square [1987: 54]), consisting of, among others, "male/female," "light/dark," "Greek/barbarian," and "human/animal." The superior first term in each case, duBois argues, aims at promoting the position of white Athenian males (1982: 4). Whereas Aristotle's example has the appearance of being chosen for its notional innocence, duBois's analysis shows that Aristotle's demonstration of the square is precisely an ideological articulation of cognition, an articulation aligned to advance specific political ends within the larger discourse of Greek culture.

This reading of Aristotle's oppositional square more precisely positions ideology within the differential, binary frames of cognition we have been examining throughout this book. Such a reading makes ideological interpretations, such as those of Jameson, more provisional and unstable, but it also enriches and expands the possibilities of discovering social "interest" in cognitive truths and of discovering the *special* situation of the seemingly general and defining case of self-evident truth. In *Science as Power*, for example, Stanley Aronowitz discusses changes in the conception of contemporary science, particularly the "blurring" of the post-Enlightenment picture of science as working out of a research-based model and as unassailable in its claim to objectivity and accuracy. Aronowitz's argument does not concern mere degradation or slippage in the institutional practice of science; rather, it maintains that the "very idea" of science is being transformed within a changing historical and ideological framework. He takes modern culture to be dominated ideologically and culturally by the economic mode of "late capitalism," a mode he characterizes as "possessive individualism," "the dominance of the market in the economy and the conduct of political affairs," and "the consolidation of bureaucratic power" (1988: 352). The immediate ideological "other," or contrast, to late capitalism for Aronowitz is the "'socialist' state," an organization of society not driven exclusively by market and entrepreneurial concerns. Aronowitz's description of democratic socialism allows him to imaginatively construct a "new science" based on, "as a condition of its emergence, an alternative rationality which would not be based on domination" (1988: 352).

Despite the language of totalization inherent in terms like "capitalism" and "socialism"—or even implicit in the sweeping general binary opposition between "domination" and its "alternative"—Aronowitz is describing the functioning of ideology in local and repeated instances, in semantic conflicts and discursive transformations. This becomes evident when his discussion of the "discourses" of different ideological and conceptual positions is mapped on Aristotle's square in terms of one of the propositions Aronowitz makes to characterize late capitalism.

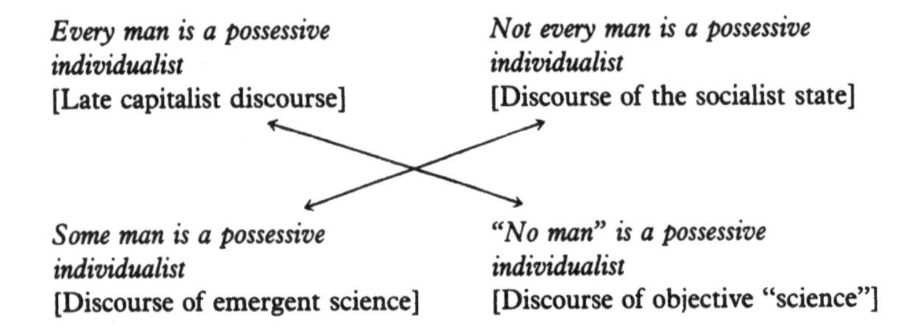

By situating Aronowitz's argument on Aristotle's square, we are emphasizing the way his argument stipulates a set of values governed by the definition of capitalism in relation to socialist ideals. In this view of science as a historical and cultural mode, Aronowitz posits the classical "discourse of Science" as a version of a rational objective management with centralized command centers that are conceived to be objectively and self-evidently necessary to maintain the "essential" hierarchic nature of knowledge. In this description, "science" as a practice directly expresses the ideological assumptions of late capitalism. Against this, Aronowitz posits an "emergent" science that doesn't simply deny the classical discourse of science, but opens a space within that discourse for alternative ("emergent") forms and modes of comprehension.

If the proposition "No man is a possessive individualist" implies that science traffics only in the general *disinterested* case, then "Some man is a possessive individualist" tells *another* story. This story suggests that parts of a complicated, complex "science" can be seen when the narrative of the special case describes the activities, and consequently the seemingly peripheral interests, of those who act. By "man," Aristotle means "human," but the human for him is so caught up in the special case of the male, that (as duBois points out) he cannot see within his knowledge the complication in his description. In the same way, the *comfort* that the old man we described at the end of Chapter 3 takes in talking to his grandson—in narrating his experiences and telling what they mean—presents a "reason" for cognition that is interested and special (as opposed to disinterested and general "reason"), namely a reason to get up every morning ("I enjoy talking to you," he had said to his grandson, "because, when you get to be ninety, you need all the reasons you can find to get up every morning"). Such reasoning offers an "alternative rationality" which does not substitute itself for traditional rationality but complicates it and allows us to ask different questions, including questions focusing on the ways culture and cognition interact. To see the emergent complication in the categories of knowing is precisely Aronowitz's definition of emergent science.

Of course, Aronowitz is also describing an idea of a mode of science that is only now emerging historically. In this way, his own definition is complicated. This complication touches on a major concern that we have elaborated from several different directions in *Culture and Cognition*. The term "emergent" encompasses the claim we have made—in Greimas's semantic (as opposed to logical) reading of narrative structures, in the mixture of chance and necessity in Darwin's account of natural history, in Lacan's descriptions of unconscious desire and unconscious cognition—that science and cognition can be *alternatively* conceived as historically situated practices that function by creating and recreating—by repeatedly *superimposing* and *narrating*—ways of knowing. What we have described in mapping and narrating the relationship between culture and cognition in some ways details the "alternative rationality" Aronowitz describes in emergent science. Through the superimpositions of semiotics, cognitive science, and psychoanalysis, we are attempting to develop a model of cognition that incorporates temporalized critique, that is capable of change (and of accounting for change), and that responds to the demands of new conceptions of the aims and techniques of disseminating and producing knowledge. In the rest of this chapter we examine new and alternative conceptions of the dissemination and production of knowledge, teaching and publication in the modern institution of learning.

Understanding Knowledge: Oppositional Pedagogy

An exemplary case of the oppositional cognitive practices of critique is the descriptions and practice of "oppositional pedagogy" as they have developed in the last two decades, the attempts by Paulo Freire, Ivan Illich, Stanley Aronowitz, Henry Giroux, Michael Apple, and others to initiate a critique and transformation of teaching within schools and universities. Their case is of particular interest in that change in education often seems to move against rigidly defined and conservative social expectations for educators and pedagogical institutions based upon the assumption that education is fundamentally nonideological. There is an almost universal fear, in other words, of the teacher-as-political advocate, of the special-pleading pedagogue who exploits a privileged position in education to satisfy ulterior political motives. In a striking gesture of such oppositionalism—related to duBois's interpretation of the logic of Aristotle's square—Jane Gallop reminds us that the term "pedagogy" carries with it specific ideological freight, reflected in the word's etymology, as the "teaching of boys." She asserts that pedagogy at base is male instruction as a metonymic connection to male social privilege (1982: 118). Within this comment is an insight that also sheds light on the connection made frequently by Sade, Dickens, and others between pedagogue and pederast—hence the universal fear of the politicization of teaching.

In this dismantling of pedagogy as a universal and timeless institution, the theorization of oppositional pedagogy unsettles the self-evident fixity of pedagogy as a singular concept and dismantles the implicit maleness

Gallop speaks of—the ideological freight at the heart of "pedagogy" as traditionally conceived in Western culture. Barbara Johnson makes a similar challenge to pedagogy by questioning the Western repression of "feminism" in teaching—the absence, in other words, of girls in the boys' school (1982: 182). She points out that the connection often exists between the flight from traditional pedagogy and the advent of a "deviant" school—in Molière's phrase—"*L'École des femmes*" (1982: 165), a practice ideologically in opposition to the traditional one. Barbara Guetti, similarly, identifies traditional pedagogy as a "male" teaching in the sense of being an unalterable practice that represses femininity (1982), a scenario she shows to be borne out in Choderlos de Laclos's "De l'education des femmes."

In these observations we can see that more than anything else, oppositional pedagogy aims to effect a fundamental reconception of what teaching is. Henceforth it cannot be taken as a value-free or ideologically innocent activity, the purity or neutrality of simple cognition to be shielded from political deviancy or contamination. It is, rather, an inherently political practice, and oppositional pedagogy shows the necessity of understanding teaching fundamentally as a situated cultural practice whose effects are "learning" only insofar as learning itself is seen in ideological terms—as having a political and cultural impact. All knowledge, in other words, does something to someone, benefits someone and oppresses someone else, empowers some and deprives others. Knowledge, as Latour claims, is an *interested* activity that exists as a dimension of actual work with effects that make a difference in the world.

Oppositional pedagogy as articulated by recent thinkers attempts to reverse the traditional, fixed hierarchy of teacher and student, active agent and passive receptacle, oppressor and oppressed, and to critique that hierarchy. This critique follows from the perception of social and cultural *contradictions* suppressed within the self-evident disinterested understanding of pedagogy. Such perception facilitates a new conception of education, not a utopian conception but one that rises from actual instances of teaching, from teaching situations as they are analyzed according to the critical connections to be made between knowledge and power. In just this oppositional style, as we will see, Gerald Graff presents actual institutional approaches to teaching—professional humanism, scientific inquiry, etc.—in the pattern of oppositional, ideological relations. He is willing, moreover, to situate himself in that discourse, articulating the relations of knowledge and power within the cultural practices and the institutional force of "English" in the United States.

Traditionally assigned a conservative function and operating within well-defined responsibilities, the pedagogue defined in such oppositional terms exemplifies the dilemma of the oppositional critic in general tied complexly to a social institution. The complexity of this situation is that such a teacher, like many of the critics we described already, is dependent on the social function educational institutions provide and yet is committed to far-reaching changes in the understanding and practice of cognitive activity within and beyond its bounds. In this *position,* oppositional pedagogues are part of the system and "play a fundamental role in producing the dominant culture," as Henry Giroux, David Shumway, Paul Smith, and James Sosnoski point out, but they also wish to offer "to students forms of oppositional discourse and critical social practices at odds with the hegemonic role of the university and the society which it supports" (1984: 480). The situation of oppositional pedagogues, like that of oppositional critics generally, is precisely that of trying to stand apart from their own institutional commitments enough to change their own cognitive activity. There is currently a good deal of attention and self-consciousness about such oppositional practices in education. Books such as Dinesh D'Souza's *Illiberal Education* (1991) have fostered an atmosphere of suspicion and distrust regarding multiculturalism, affirmative action, cultural studies, and almost all attempts to revise curricula. This lumping of practices that supposedly represent academic irresponsibility, the alternative charges of political leftism and simple posturing, and the general distrust of university faculty draw attention away from any serious critique of education or the curriculum.

That critique has been articulated in recent theories of teaching and the institutional functioning of teaching that suggest that the difficulties of oppositional pedagogy persist at the heart of what "pedagogy" already is in its traditional conception, at the core of what Freud called—along with government and psychoanalysis—the "impossible profession." The recent and still most influential text advancing a revolutionary pedagogy is Paulo Freire's *Pedagogy of the Oppressed* (1968). This book made a case for a dissenting pedagogy based on political involvement by those in the academy who do not want merely to understand but to change education and the inculcation of cognitive practices more generally. Freire's theory is, therefore, a critique but also a plan to change the institution of education in light of three principles, all of which could be said to constitute oppositional pedagogy. First, Freire looks at the way school curriculum and institutional practices often inadvertently collaborate with and in various ways support objectionable governments and social systems. In response to this, he advocates the development of what he calls *conscientizacao,* which he defines as the "ability to perceive social, political, and economic contradictions" in society and culture (1982: 19). Not a simple lesson to be learned or taught, or an "answer" to a problem, *conscientizacao* is a perspective that deepens as it persists. In this formulation "contradiction," of course, refers to class conflict and the critique of such conflict as a primary insight on which to base all other instruction. But as we will see, it takes the shape of the kind of cultural cognitive contradiction we found in Aristotle. For Freire, discerning the implications of the contradictions of class conflict begins but is also the ultimate aim of a student's education.

Freire's second principle regards the active participation of students as the oppressed struggling in their own liberation/education. The "oppressed," he says, "must be among the developers of this [revolutionary] pedagogy" (1982: 39) as they become active in the events usually reserved for teachers and administrators. The degree to which they participate reduces their alienation and, in a sense, is the real goal of their liberation. Once they know they can participate in education, as Freire argues, students do not learn passively by allowing pedagogical decisions to be made by others—administrators and "experts" not themselves subject to the teaching they supervise. Rather, students thereafter produce knowledge themselves through participation in teaching and the making of institutional decisions about what will be taught and who will teach.

The third principle marks the extent that change must reach in education to have a lasting effect. Freire stipulates that the "historical task [or ethical responsibility] of the oppressed [is] to liberate themselves *and* their oppressors" (1982: 28, emphasis added). Beginning with *conscientizacao,* this process is complete when pedagogical practices are altered and transformed in order to be more socially responsive. Students must see themselves differently for cultural change to occur, become responsible for their own education, but the relations constituting education and the "dissemination" of knowledge must also be reconstituted. Whereas in the traditional scene of instruction—as Giroux, Shumway, Smith, and Sosnoski describe it—"experts in a discipline impart to apprentices the received knowledge about a particular subject matter" (1984: 481), the teacher being the active subject imparting knowledge to passive students, in the new pedagogy students and teachers exchange active and passive roles and even alternate being teacher and student. In this way, as Freire points out, students are liberated from the role in which they are "alienated like the slave in the Hegelian dialectic" (1982: 59), and the teachers, too, are liberated from the equally alienating "master" role in hegemonic instruction. This change allows "both parties to construe themselves as agents in the process of their own cultural formation" (Giroux et al. 1984: 482).

Often referring to these precepts of a "revolutionary" pedagogy, oppositional critics since 1968 have extended Freire's analysis to specific curricular concerns in the teaching of reading and writing, history, business, and other areas, and they have also tended toward even more radical and sweeping analyses of modern education (see Apple 1982a; Apple and Weis 1983; Illich 1970). Other recent works have continued to focus the radical critique of education in line with broad political and social concerns (see Aronowitz and Giroux 1985; Freire 1985; and Livingstone 1983). A pointed and useful analysis of the theory of oppositional pedagogy is *Criticism and Social Change* (1983), in which Frank Lentricchia cites John Dewey and Kenneth Burke—both of whom foreshadow Freire—to reintroduce Freire's distinction between "education as a function of society" and "society as a function of education." These two conceptions of education, Lentricchia notes bluntly, divide "the world between those who like it and those who do not" (1983: 1). Those adopting the former view generally imagine education to have an inculcating function in society—to be a kind of training, or ascesis, in the rigor of accommodating and joining an already constituted circle, the community of knowing subjects whose knowledge is objective and static. Those of the latter view imagine education to be ongoing in a radical sense as society itself and cognitive activities emerge into existence. In this view education can be but an approximate (indeterminate) interpretation of changing social institutions—including institutions of cognition—in terms of a formulated but evolving critique. Thus, as Dewey, Burke, and Freire believe, education is fundamentally "oppositional" (evolving in opposition to what came before) because it represents culture's continual evolution.

The concept of critique—again Freire's *conscientizacao* and essential to oppositional pedagogy—is largely oppositional in the sense we have already developed. In this context, as we have seen, any pedagogical event (a classroom lesson), like any cognitive event, is ideologically motivated and serves the interest of certain groups and not others. In this view of pedagogy as *interested,* Freire wishes to theorize the specifically functional value of resistance that informs a historical discourse in particular ways at particular moments in cultural life. Biological evolutionism, for instance, might be foregrounded as instruction at one moment in history—as it was during the hegemony of laissez-faire capitalism—and that choice and the panoply of rejected choices it entails (biblical creationism, Native American creation stories, etc.) would then constitute a system of designated and canceled options, a system of resistances facilitating some and blocking other subsequent historical developments. Or methodological binarity might govern research and understanding in a time of social upheaval and articulated challenges to long-held belief. Such a functional approach to cultural cognition as constituted by a kind of ideological "discourse," in which some possibilities are structurally allowed and others disallowed at particular moments, is what Marxists often aim for when analyzing cultural events in terms of historical causation, what Jameson calls an "absent cause." It is articulated in the oppositional square we have examined.

Even more than Aristotle's square (or rather duBois's reading of it), oppositional pedagogy attempts to account for its own institutional placement, showing, as Jim Merod says, an "understanding of the institutional context within which students and teachers do their work—from which they can see their work as interpreters to be situated with concrete social and political consequences" (1987: 151). It is on this basis that Merod criticizes Jameson's readings of literary texts we examined earlier for the way they articulate ideology but are "unavailable [to show] the passage from interpretive theory to the making of a body of critical intellectuals who know where they work, what they work for, and how they mean to achieve regenerative social goals" (1987: 18). Teachers and critics, as Merod argues, need to take the measure of their own "insertion" within "the divisions of labor that continue to define our economic and social world" (1987: 19). This means theorizing the political and social context of their work as teachers and intellectuals—their cognitive activity—which takes into account the social impact of their labor and what they want to achieve. They cannot, as Merod accuses Jameson of doing, leave "the creation of a critical [and pedagogical] community to one side as a later, more revolutionary moment of theory" (1987: 18).

But if Jameson errs by taking for granted the institutional context of pedagogical work, oppositional pedagogues who try to situate their work must also contend with the formidable problematics, some say the impossibility, of generating a critique and working toward change within an existing institution that has already established its aims and means as seemingly self-evident truths. A case in point is the critique advanced by members of The Group for Research on the Institutionalization and Professionalization of Literary Study (GRIP). In "The Need for Cultural Studies: Resisting Intellectuals and Oppositional Public Spheres," the GRIP critics (Henry Giroux, David Shumway, Paul Smith, and James Sosnoski) advocate the rise of cultural studies as an oppositional pedagogical practice for American society. They reason that the proliferation since the late 1960s of interdisciplinary programs in American universities was initially a practice conceived to counter the reigning ideologies that govern the exercise of power in American life—"both the enabling and constraining dimensions of culture" (1984: 473). The cultural critique emanating from such enterprises, they argue, is needed "to identify the fissures in the ideologies of the dominant culture" (1984: 473). Gradually, however, these programs "succeeded" to the point that they became "institutionalized" as socially ensconced entities and inevitably began to harmonize more with institutional expectations concerning the social problems they isolated and the conclusions they drew. Finally, in the 1980s these programs were institutionally assimilated to the point where they lost their oppositional force and had no critique to advance.

The GRIP critics see this failure of interdisciplinary programs as mandating a whole new oppositional strategy. Whereas oppositional critics generally imagine oppositional practices to be situated *within* institutions, the GRIP critics next reason that there can be no exchange between oppositional forces and the institutions they wish to supplant or change. In other words, for them the possibility of assimilation and ideological contamination dictates moving intellectual work completely outside existing institutions. Their analysis of the disappointing "success" of interdisciplinary programs in American colleges and universities leads them to suggest that in the future none of these cultural studies programs can be housed in existing universities or research institutes where they will be professionally assimilated and, thus, contaminated. Rather, they seek a proliferation of "new" cultural studies institutes committed to avoiding the ideological pitfalls of existing organizations. These institutes, virtually noninstitutes, will start on fresh (noncontaminated) terrain, manage to transcend professional compromises, and generally avoid the ideological boundaries that traditionally limit the universities.

These speculations are no doubt useful and fascinating as utopian projections, but they do not constitute a critique of pedagogical work and to that extent are disappointing as an attempt to situate the work of cultural criticism. That is, whereas Jameson avoids institutional questions by having them disappear within texts, the GRIP critics project auxiliary cultural studies institutes as utopian possibilities. There probably is a need for new research institutes in the United States, but while the GRIP critics promise and begin to provide an analysis of the institutional context for oppositional pedagogy, they finally are too little interested in the specific context of work currently being done and, like Jameson, are "unavailable" to explain "the passage from interpretive theory to the making of a body of critical intellectuals who know where they work, what they work for, and how they mean to achieve regenerative social goals" (Merod 1987: 18). Instead, they escape the resistance of textual and institutional practice too well, are too eager in their utopianism, and disappear into the future, fictive institutes they have projected as pedagogical ideals. In short, neither Jameson nor the GRIP critics illuminate the logic of cultural opposition, the system of intellectual work, that connects (and disconnects) them to the institutional site of their own intellectual labor. In the absence of a critique of the logic of opposition itself, oppositional pedagogy remains ineffectual, a source of detached observations about teaching and social change.

In his recent work Gerald Graff comes close to providing the institutional critique needed, at least in one discipline, to understand oppositional pedagogy as a cultural force. In *Professing Literature: An Institutional History* (1987), Graff describes the institutional discourse that has taken place in the United States since 1828 as "English" has become a professional discipline—both a department in universities and colleges, and a body of "expert" professional knowledge. Graff shows how American universities at first imitated German models in the teaching of philology and language and then gradually developed an "English" curriculum with the advent of graduate education at Johns Hopkins University, Indiana University, Cornell, and Princeton. Graff's "institutional history" is an account of intensified professionalization and the developing role of "expertise" in the American academy, the movement of the university away from clerical training and ancient language study and toward scientific and critical inquiry within well demarcated professional boundaries. Graff relates the history of how American universities have achieved the social prestige and cultural impact they have today.

Graff's complex historical account identifies four pedagogical positions that interact with one another within the American academy. That is, both in terms of its historical development and its current makeup, the "profession" of English studies can be divided into four areas of activity or inquiry. The most pervasive, and in many ways still very influential, Graff calls professional "humanism," by which he means an institutionalized belief that there is a "truth" content of human affairs communicated through the liberal and systematic inquiry into English and American literature. The underlying, romantic assumption here is not only that an essence of human nature underlies humanistic activity but that the humanities, properly understood, naturally reflect this essence and when unencumbered naturally bring it forth.

Once the English department in the United States became a professional guild, this professional humanism became the underlying, guiding orientation of the "field-coverage" system of departmental organization. As Graff explains, "for reasons having to do equally with ensuring humanistic breadth and facilitating specialized research, the literature department adopted the assumption that it would consider itself respectably staffed once it had amassed instructors competent to 'cover' a more or less balanced spread of literary periods and genres, with a scattering of themes and special topics" (1987: 6–7). This organizational format even now "seems so innocuous as to be hardly worth looking at" (1987: 7), and yet it expresses a "faith that exposure to a more or less balanced array of periods, genres, and themes would add up in the mind of the student to an appreciation of humanism and the cultural tradition" (1987: 9). Underlying this humanistic enterprise is the belief that "*literature teaches itself*" (1987: 9–10), and, thus, there can be little need to discuss the theoretical adequacy of the "field-coverage" scheme or speculative issues related to it. This is true, Graff concludes, because implicit in such a system taken to be natural and inevitable is "the illusion . . . that nobody [has] a theory" (1987: 9).

Prior to professional humanism in much of the nineteenth century there existed, and still persists, an approach to literary study that can be called appreciation of the "classical college." This approach originally developed in an educational system in which the basic mission was preparation for professional study, "Christian leadership and the ministry" as well as medicine and law (1987: 20). As a "liberal studies" education, it was concerned with "gentle breeding," "appreciation," and "acculturation for 'the cultivated gentleman'" (1987: 20). It originally centered on the study of Greek and Latin so as "to inspire the student with the nobility of . . . cultural heritage" (1987: 28). This education often focused on the grammar of these languages, and it was believed that the wisdom of Greek literature "would somehow rub off on students through contact with linguistic [and grammatical] technicalities" (1987: 35).

While this approach is little evident today as a professed curriculum, it exists as a residue in many English departments. It is the official orientation of a very few, and yet it is the commitment of considerable numbers of individual professors. This is the viewpoint of those who disapprove of the entire direction of modern literary studies, including its emphasis on research and publication, and who still imagine returning to an "innocent"—or at least not so professionally involved—appreciation of literature based on the natural and pure love of words. There are strong overtones of this "appreciation" as an approach, for instance, in Allan Bloom's *Closing of the American Mind* (1987), which promotes mental development, the shaping of mind, through close reading of the classics (almost every English or modern language department has one or two stalwarts of this classical-college persuasion). While not a dominant approach to literary studies, it provides a strong and helpful contrast to the more professionally oriented, and traditionally dominant, institutional humanism.

The third major approach to literary studies is that of scientific and rigorous methodology, which Graff traces to the nineteenth-century "German-trained cadre of scholarly 'investigators,' who promoted the idea of scientific research and the philological study of the modern languages" (1987: 55). Beginning with—and in some sense always tied to—nineteenth-century philology, the new academic professional of the late nineteenth and early twentieth century was an expert working out of established models of discovery aiming to add fundamentally to the store of knowledge, this professional's loyalties going, as Graff says, "to his 'field' rather than to the classroom dedication that had made the older type of college teacher seem a mere schoolmaster" (1987: 62).

In the early twentieth century, this new scientific professional in the humanities soon became diversified in literary historical scholarship and other closely related areas not specifically philological. Eventually, with the rise of the New Criticism in the 1930s, the new "scientific" interest was carried over to speculative concerns with imagery, rhetoric, and aesthetic topics such as paradox, irony, ambiguity, and symbolism. This development toward criticism per se conflicted from the start with research-based (historical and linguistic) activities as speculation against received notions, a conflict that continues today in the oppositions between theory and scholarship, and between theory and humanism. In the broader sense, though, through its rational principles, its pragmatic and teachable methods, and the demonstrable results it achieved, the New Criticism of this period continued the move toward scientific rigor and professional expertise and away from the "appreciation"-based or even humanistic concerns of the earlier approaches.

The fourth approach is the one that Graff's own work represents in its attempt to situate the teaching of English as a component in a complex of cognitive-cultural activities. To do this, Graff endeavors to read the discourse of English (as a professional activity) that has been evolving since the early part of the nineteenth century in the United States. He identifies pedagogical practices that have moved to the foreground to dominate the profession and also the way in which these practices have caused the English profession to transform itself as an entity in the academy. Moreover, he looks at the impact of these practices on the immediate university work environment and on the areas of culture they have the strongest impact on. He shows that each development in "English" is overdetermined as it arises through a series of interactions across and among cultural spheres. His concern always is to identify lines of force and resistance that expose the ideological pattern of the English department's formation. The "resistances" identified in each approach mark the boundaries of what can be taught or even imagined at any moment within "English." Achieved in Graff's analysis is not so much a rendering of how "English" has taken shape as a discipline in the academy as a critique of what has happened in culture—focusing on the connections, as Aronowitz and Giroux say, between power and knowledge—so that "English" emerged with the cultural impact and professional shape it has.

In such an analysis Graff has adopted the underlying assumptions and procedures of the oppositional critic. He does this, in part, as he positions his own activity, the critic's work, as a practice in relation to other real and potential practices, that is, within a specific ideological and work environment. It is easy, finally, to place Graff's own work within the very professional discourse he discusses, within the logic of pedagogical practices. His is precisely the practice of oppositional pedagogy that Jim Merod called for. In doing this work, in other words, Graff takes up the *position* left vacant by the GRIP critics in their utopianism and by Jameson and his critique of narrative from the vantage of objective cognition.

Most fundamental to Graff's work, though, is the "oppositional" strategy of his critique, the way in which he attempts to map the "logic" of cultural resistance working through the practices of pedagogical discourse by assuming the common foundation for cognition and ideology we examined earlier. For this reason, his analysis of English studies in the United States can be inscribed on a oppositional square. On the first level there is a coupling of propositions that form a "contrary" relationship in Graff's opposition between professional humanism and "appreciation" in the classical college. These two terms oppose each other arbitrarily to mark a range of difference. By "arbitrarily," we mean that the coupling of these terms begins with the supposedly self-evident truth of humanism as an object of knowledge. Against this arbitrary beginning, however, the opposition establishes a range of cultural references, a spectrum of possible significations, as in humanistic appreciation, classical humanism, and so on. (For discussions of the New Humanism, transpersonal humanism, and impersonal humanism, see Davis and Schleifer 1991: 55–68.)

On the second level the third term indicates the specific hierarchy, ultimately the ideology, by which meanings are being organized within the range of reference established on the first level. As we have said, the third term, in effect, "interprets" the first level by describing the "nature" of the category upon which the first-level binary opposition is inscribed (see Schleifer 1987a: 54). In Graff's opposition between professional humanism and "appreciation," the third term is "rational (scientific) inquiry" as the designation of power in the professionalism/appreciation opposition, the signal that rationality will be the controlling term. Scientific investigation as indicator of ideology in this oppositional square, therefore, "understands" the first level of reference, interprets and limits it, by identifying the hierarchy that organizes it (the very hierarchy elevating rationality over other forms of apprehension with which we might begin). The third term, thus, inscribes a set of values positioned as a particular pattern in the square's hierarchy. Graff could have as easily described this third term as "cognitive objectivity" or "objective understanding." In this term cognition is *instituted* as the abstract objective ground (such as Whitehead describes in mathematics) of self-evident truths.

The fourth term then completes this ideological reading with "oppositional pedagogy"—the reimagining of the relationship of power and knowledge (with the focus on the elements of power within "knowledge") in education we described earlier. At this farthest reach of the square, "oppositional pedagogy" contradicts the position of "appreciation" and stands in opposition to "scientific research." In this, it gives final expression to (even while it challenges) the rational humanism from which this cognitive schema for "English studies" is generated. This fourth term, in effect, goes "beyond" (as if to "escape," or to "exceed") the other oppositional relations. Whereas the first-level relationship signals the coexistence of "contrary" possibilities, and the third position articulates the values inherent in the square, the fourth term is a potential resituating of this discourse—what we have seen Lacan call in a similar context a resituated subjectivity, a new *moi*. Once again, the fourth term foregrounds an "other" potentiality within rational cognition—the potential of critique—made possible by the play of difference between the initial terms humanism/appreciation on the first level. In this way, the fourth term marks a new authority emanating from but, at the same time, alien to the self-evident truth it participates in.

The economy of Graff's humanism/appreciation opposition can be projected as follows:

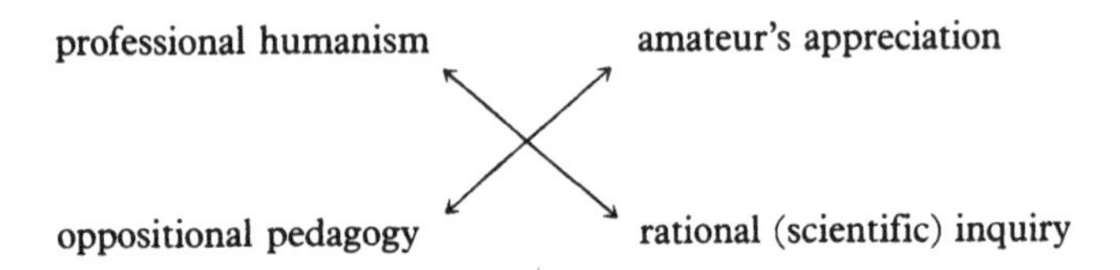

These relationships are all established by an oppositional tie, of one kind or another, to the first term, except in the case of the fourth term, which "closes" the square by articulating a possibility effaced, or repressed, in the articulation of the first three terms and, in so doing, recognizes the possibility of a new pedagogy. Oppositional pedagogy comes into the square (comes into "mind") as the instance of what was suppressed all along through the insistence in the discourse on various formal oppositions. The square, then, implicitly recognizes the fourth term as the completion but also as the potential destruction—the oppositional "Other"—of its own cognitive activity. In this way, the oppositional square shows "oppositional pedagogy" to be a historical phenomenon, a specific way of thinking emerging as a "rupture" that has left its trace but is not yet representable as a specific cognitive practice. This is precisely Freire's conception, too, of the "pedagogy of the oppressed" as being, at this historical moment, within the articulations of the dominant bourgeois thinking, "without a discourse" and, at the same time, on the verge of emerging, an emergent but not yet recognizable pedagogical cognitive activity. As such, the power of such pedagogy persists within the academy specifically as an effacement: hence the disruptive nature, the scandal, of the "political pedagogue" as an agency of power. The representation of this double writing, of teaching as a neutral and subversive activity, is at the heart of the relationship between knowledge and ideology.

It is in this way that Graff's reading of the opposition between humanism and "appreciation" demonstrates the economies of ideology engendered by binarity, specifically the economy of politicized teaching. The result is a situated pedagogical discourse inscribed by the square's four terms—themselves rising from and transcending the unintelligibility of "mere" differences, unorganized and nonpoliticized heterogeneity. That is, what we are viewing in Graff's method of critique, as with Jameson, Said, Greimas, and the oppositional critics generally, is an ideological approach to the problematics of teaching. Ideology, in this view, is not an immutable system of belief but an "effect" of interested cognitive activity produced within a system of differences that are continually subject to transformation within that system. One cannot arrest the cycle of cognitive-semiotic activity within ideology because its momentum is tied to history. As a cultural phenomenon, it is a manifestation of "history"—the diachronic emergence of the unthought and the initially unreadable. Such emergence takes the form, as Julia Kristeva says, of a new conception of ethics, ethics conceived as the cognitive and cultural energy released at the moment when the "unthought" becomes thinkable, at the moment of cognition. For English studies, this is the moment which Graff traces in his book, when the publication and dissemination of literary studies became a legitimate articulation of knowledge.

Producing Knowledge: The Ethics of Publishing

John Dewey, in his *Outlines of a Critical Theory of Ethics* (1891), characterizes "ethics" as a science, or a mode of inquiry, and he compares ethical inquiry to other sciences that study human action: anthropology, psychology, and sociology. The difference between ethics and these other sciences is that "these latter branches of knowledge simply *describe* while the business of ethics is to *judge*" (1969: 241). For Dewey, then, ethics is a cognitive activity, just as for Jameson, political critique is a cognitive activity. Dewey does not mean that ethical inquiry is prescriptive. Instead, what distinguishes ethics is the nature of its inquiry, that is, the type of questions it poses: "its business is to detect the element of obligation in conduct, to examine conduct to see what gives it *worth*. Anthropology, etc., do not take into account the *whole* of the action, but simply some of its aspects—either external or internal. Ethics deals with conduct in its entirety, with reference, that is to what makes it conduct, its *ends*, its real meaning" (1969: 241). The human sciences, for Dewey, are inadequate in that their mode of inquiry omits the question of the "end" of human action and consequently misses the "real meaning" of human conduct. Ethical inquiry, however, asks the question of "end," which Dewey equates with the question of "worth" and which we have just equated with emergence in time. Editors of scholarly publications are precisely situated to govern the emergence of knowledge—the institution of knowledge—and the ethics of publishing is the inquiry into this situation.

Responding to the Continental philosophical tradition rather than the Anglo-American tradition, but writing only four years earlier than Dewey's *Outlines,* Nietzsche placed a similar value on ethical inquiry as the basis for the human sciences. In the note that concludes the first essay of *On the Genealogy of Morals* (1887), Nietzsche extends his project of transforming Kant's critical philosophy to all sciences: "*All* the sciences have from now on to prepare the way for the future task of the philosopher," Nietzsche writes; "this task [is] understood as the solution of the *problem of value,* the determination of the *order of rank among values*" (1967: 56).

In extending ethical inquiry to the sciences, Nietzsche hopes to direct the study of values to the sciences themselves, to the practices that produce knowledge and to those who are engaged in those practices. The great problem of knowledge, as Nietzsche remarks in the opening sentences of the *Genealogy,* is that "we are unknown to ourselves, we men of knowledge—and with good reason. We have never sought ourselves—how could it happen that we should ever *find* ourselves?" (1967: 15). Nietzsche's project—which he calls a genealogy and which has been revived in more recent times by Gilles Deleuze and, especially, Michel Foucault—is not simply an ethical inquiry, that is, a mode of inquiry focusing on the values that underwrite human conduct. In addition, and primarily, Nietzsche's project is inquiry as responsibility, focusing on the very cognitive activity that generates "underwritten" value. In other words, it responds to what Julia Kristeva calls an overwhelming, "shattering" experience. "The issue of ethics crops up," she writes, "wherever a code (mores, social contract) must be shattered" (1980: 23). It *crops up* in relation (and response) to cognitive disorientation.

In a recent essay entitled "The Search for Grounds in Literary Study," Hillis Miller begins by describing reading itself as disorienting, uncanny, comparable to what George Eliot narrates as "sudden, inexplicable fits of hysterical terror or of 'spiritual dread'" (1985: 19). Miller compares reading to this passage from *Daniel Deronda* and another from Maurice Blanchot "quite arbitrarily," as he says, "or almost quite arbitrarily, as parables for the terror or dread readers may experience when they confront a text which seems irreducibly strange, inexplicable, perhaps even mad" (1985: 19). "On the one hand," Miller goes on,

> a good reader . . . especially notices oddnesses, gaps, anacoluthons, non sequiturs, apparently irrelevant details, in short, all the marks of the inexplicable, all the marks of the unaccountable, perhaps of the mad, in a text. On the other hand, the reader's task is to reduce the inexplicable to the explicable, to find its reason, its law, its ground, to make the mad sane. The task of the reader, it will be seen, is not too different from the task of the psychoanalyst. (1985: 20–21)

Miller begins the search for the grounds of the cognitive activity of literary study with parables of reading because his is above all an ethical enterprise, an inquiry into conduct. That conduct is the cognitive activity of an editor, who, after all, is a reader par excellence. Both reader and editor "reduce," as Miller says, experience to explanation. Moreover, even if the task of the scholarly editor is quite a bit different from that of the psychoanalyst, the ethics of these tasks—their conduct—is much the same. Both editor and psychoanalyst are figures of the law, authorizing texts, piecing them together, even if in Freud's talking cure the psychoanalyst is an editor-in-reverse who returns (i.e., "submits") the patient's language to him or her as a kind of published version to be heard in an unedited form. In Miller's parable the psychoanalyst as reader grounds and authorizes discourse in his "publication" of the patient's speech in the same way that the scholarly editor is faced with the ethical choice of grounding the texts that come across his or her desk.

Still, the psychoanalyst's task is quite a bit different from the editor's because the psychoanalyst seeks to make the patient a *writerly* reader, in Roland Barthes's terms (1974: 3–4)—to make the patient attend to the oddness of his or her discourse—while the editor attempts in reading to discover reason, law, and the ground of understanding. The editor grounds his or her decision on the degree of the domestication of discourse. In an important way Barthes's concept of "writerly" discourse attempts to situate the reader as the psychoanalyst's patient finding his or her discourse in the discourse of the Other (and finding the discourse of the Other within his or her discourse), quite literally to *authorize* the reader—to make him or her an "author"—whereas the editor attempts to authorize a *readerly* writing, a discourse, like that of the psychoanalyst, that is not his or her own. But of course in this, too, and again quite literally, the editor makes the writer an "author."

In his essay Miller mentions four "grounds" of literary study, social, psychological, linguistic, and metaphysical, and it is noteworthy that we have already made reference to the first three in the figures of editor, psychoanalyst, and literary theorist. These grounds can be placed on a semiotic square, beginning with a psychological (or "subjective") ground opposed to a linguistic (or "structural") ground. A third ground is social, the kind of "oppositional" stance we have just examined. In a way, we have been traversing these grounds—the "mind" of cognitive science, the "sign" of semiotics, the social "situation" of cognitive activity—throughout this book.

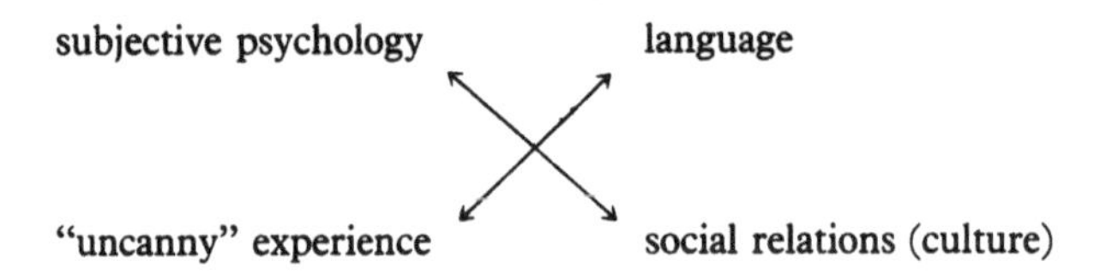

The fourth category or ground Miller describes comes closest to the enigma to the mind Jameson sees in the fourth position of the square. Miller describes it as the *uncanny* experience of reading with which he begins his essay, and it is closest to the cognitive elements in ethics we want to examine here. Miller asserts that the "fourth possibility for the disturber of narrative sanity and coherence, a disruptive energy [that is] neither society nor individual psychology, nor language itself is properly religious, metaphysical, or ontological, though hardly in a traditional or conventional way" (1985: 21). He follows Blanchot in calling this possibility "ontology without ontology," and later he follows Derrida in suggesting that it is that which disrupts the principle of reason (1985: 21, 29). Above all, he says, it is "something in our relations to other people, especially relations involving love, betrayal, and that ultimate betrayal by the other of our love for him or her, the death of the other" (1985: 21). This category, then, is the most closely related to ethics itself in its dimension that is not simply social or psychological or linguistic but that, in its way, encompasses all of these in terms of what Dewey calls "inquiry," Nietzsche calls the transvaluation of all values, and Miller calls "'criticism' in the fundamental sense of 'critique,' discriminating testing out . . . between theory and practice" (1985: 30–31). In its negativity—negating social relations, psychological states, the possibilities of discourse—the fourth "ground" comes close to what Barthes describes as the "movement" of his work as a whole in what seems to us to be strikingly *ethical* terms, "*a tactics without strategy*" (1977: 172).

That this is "ethical" can be seen in the semiotic terms by which Julia

Kristeva defines the ethical (terms, we should add, that are hardly traditional or conventional). "Ethics," Kristeva writes, "used to be a coercive, customary manner of ensuring the cohesiveness of a particular group through the repetition of a code—a more or less accepted apologue. Now, however, the issue of ethics crops up wherever a code (mores, social contract) must be shattered in order to give way to the free play of negativity, need, desire, pleasure, and jouissance, before being put together again" (1980: 23). Kristeva's ethics, like Miller's fourth "ground" of literary study, is an enabling negativity, something that creates the possibility of heterogeneity, strangeness, bewilderment, loss. It is the problematic nonreducible opposition we mentioned earlier between the special and the general case, between interested and disinterested cognitive activities. "You have learnt something," Undershaft tells his daughter in *Major Barbara;* "that always feels at first as if you have lost something." Yet here, above all, in learning and knowledge—pedagogy and cognition—conceived as negativity and loss, ethics crops up as an inquiry into conduct and the determination of a hierarchy of value. Here, above all, ethics has to do with the power and conduct of the grounding authority of cognition.

The ethics of the determination of knowledge—of both editorial and psychoanalytical conduct—is, indeed, another story from its parabolic significance in Miller. It is a "story" of power, the unarticulated "story" of the ideology of academic understanding: its ambiguous authorization of "knowledge," the grounding of literary study in authorized, that is, published, discourse. In *The Discourse on Language* Michel Foucault says, "I am supposing that in every society the production of discourse is at once controlled, selected, organized and redistributed according to a certain number of procedures, whose role is to avert its power and its dangers, to cope with chance events, to evade its ponderous, awesome materiality" (1972b: 217). Foucault identifies three general types of control over the production of discourse which correspond, more or less in material conduct, to the first three "grounds" of literary study Miller articulates. The first two he characterizes as "external" and "internal." The external type of control takes the form of social *exclusion;* that is, society exercises broad principles of exclusion that determine *what* can be written (the exercise of prohibition or censorship); *who* has the general capacity to write (the division between rational and irrational); and a system of exclusion based on what Foucault calls "the will to truth," a will grounded in the belief that has guided philosophy since Plato: "the highest truth no longer resided in what discourse *was,* nor in what it *did;* it lay in what was *said*" (1972b: 218).

The "internal" type of control over the production of discourse involves "rules concerned with the principles of classification, ordering, and distribution" (1972: 220). Foucault identifies three principles, or concepts, that are used to give order to discourse from the inside of discourse: "hierarchy," "author," and "intellectual disciplines." It is the concept of discipline that Foucault develops most fully (even if it is the concept of author that most closely approximates Miller's psychological ground of literary study). "In a discipline," Foucault writes, ". . . what is supposed at the point of departure is not some meaning which must be rediscovered [i.e., the function of "commentary"], nor an identity to be reiterated [i.e., an author]; it is that which is required for the construction of new statements. For a discipline to exist, there must be the possibility of formulating—and doing so ad infinitum—fresh propositions" (1972b: 223). For a new proposition to be accepted, however, it must "fulfil some onerous and complex conditions before it can be admitted within a discipline" (1972b: 224). Foucault identifies these conditions in terms of three requirements: the proposition must refer to a specific range of objects; it must utilize well-defined conceptual instruments and techniques; and it must fit into a certain type of theoretical field. These disciplinary conditions "constitute a system of control in the production of discourse, fixing its limits through the action of an identity taking the form of a permanent reactivation of the rules" (1972b: 224). Such internal "rules" come close to the structures that, Freud and Lacan argue, function impersonally in our psychological lives.

The third general type of control over the production of discourse involves "the conditions under which it [discourse] may be employed, and thus denying access to everyone else" (1972b: 224). This type of control is what Foucault calls, in *The Archaeology of Knowledge,* the "enunciative modality" (1972a: 50–55, 88–105). Such a modality grounds discourse in the linguistic function most broadly conceived, in language as particular instances of enunciation. In his descriptions of such enunciative conditions in *The Discourse on Language,* Foucault takes into account systems of publication and distribution. One type of system he describes comes closest to scholarly publishing. It is the "fellowship of discourse, whose function is to preserve or to reproduce discourse, but in order that it should circulate within a closed community, according to strict regulations, without those in possession being dispossessed by the very distribution" (1972b: 225). Another system of distribution he identifies is what he calls "the social appropriation of discourse." He offers education as an example: "Education may well be, as of right, the instrument whereby every individual, in a society like our own, can gain access to any kind of discourse. But we all know that in its distribution, in what it permits and what it prevents, it follows the well-trodden battle-lines of social conflict. Every educational system is a political means of maintaining or of modifying the appropriation of discourse, with the knowledge and the powers it carries with it" (1972b: 227). In his examination of what *The Discourse on Language* calls "the order of discourse," Foucault is exploring the grounds of the order of publishing, the ethics of publishing in the sense of ethics as inquiry into the "end," taking "into account the *whole* of action," including the historical situation of publishing. Thus, scholarly publishing as a discursive and political practice within institutions that sustain it and are, in turn, sustained by it can be understood within this tripartite framework of external, internal, and discursive controls—just as its cognitive "conduct" can be understood as governed by the more abstract grounds of literary study Miller articulates.

Joel Conarroe articulates a narrative of one form of such discursive "control" in describing the editorial processes of the *Publication of the Modern Language Association* a few months after he began editing that journal in 1978. After narrating the elaborate procedures of editorial evaluation for *PMLA*, he adds that "the best essays we receive would probably be accepted even if there were no specialist readers, no advisory committee, and no editorial board" (1982: 216). This narrative is remarkable in its effacement of the editor. Here, scholarly writing speaks for itself, with full, unmediated authority, in no need of an editorial subject. Or rather, "truth" speaks for itself, determining who can write and what can be written, and Conarroe—in a long line of *PMLA* editors—articulates what the anonymous "editorial policy" of the journal has for many years stated as its goal, the publication of work that "is excellent and likely to be of permanent value" (1964), what a recent editorial statement stills calls "the best of its kind" (1986).

Conarroe's narration of the author's authority is both fanciful and playful, but it articulates the *conduct* of the editor as a *conduit*, a facilitator of external and transcendental "knowledge." In this story the editor disappears, transformed to the passive acceptance (and, of course, the passive recognition) of "the best essays." Conarroe's description of the editor as simply another member of the editorial board, what he calls the "galaxy of editors," reinforces this point.

A different story of editing describes the editor as author, not simply the final authority but the single one, embodying a personal "vision" in his editorial work which the writers he edits and publishes facilitate in the same way the psychoanalyst facilitates the patient's achievement of personal, psychological authority. In a retrospective account of the first decade of *New Literary History,* Ralph Cohen tells another story. "In writing about *New Literary History* and the vision that I had of it," he notes, "I feel the embarrassment of a strange autobiographical venture. The journal involves my vision, but it depends on the work of others to fulfil it. The others are the authors of the manuscripts, my colleagues and readers who comment on the manuscripts, and the audience to which the journal is addressed" (1982: 254). Cohen goes on to ask why one would "seek to present a vision in a journal when one can write a book? Why create a vision dependent on others when one can exercise almost complete control?" (1982: 254). Cohen answers these questions in terms of the quality of his "vision" that governs editing. "Communal values have not been what my vision entails," he notes. Rather, "the challenge of the journal has been in the questioning of received views and in the testing of fashionable ones. It has offered explorations of subjects little known and seldom considered. And although each issue is devoted to a single subject, no subject is exhausted by a single issue" (1982: 255). Here, rather than reducing the editor to a passive acceptor of the best, of knowledge conceived as self-evident and transcendental, Cohen rationalizes the conduct—and the element of obligation in conduct—in terms of a program of knowledge. His story transforms the "best" into the "appropriate."

While diametrically opposed—they exist in a contrary binary opposition—both of these narratives of editorial activity imply what Foucault calls the will to truth. They both imply a sense of publishing as the authorization of transcendental knowledge, whether that knowledge is conceived as communally or individually recognizable. Against such definitions of the objects of publication as truth, ethics can concern itself with the *situation* of publishing, its power rather than its knowledge, how it sustains discursive frameworks of understanding. In this regard, it looks to the contradictory of the conception of objective, transcendental truth Conarroe articulates. Such an inquiry will look into the statements of editorial policy as occasioned *enunciations,* governed by and responding to the historical moment of their articulation. The clearest articulation of such an enunciation—that is, of an editorial statement of responsibility as a historical response—is the first such American statement, B. L. Gildersleeve's note to the first volume of *American Journal of Philology* published in 1880. (This was the first American *institutional* journal in the humanities.) Encouraged by many "prominent" colleagues, Gildersleeve writes, "I made arrangements for the printing of the Journal, and though the appearance of it has been somewhat delayed by the intervention of vacation . . . , still the Journal comes out within a reasonable time after the announcement of the project" (1880: 2). The "reasonable time" Gildersleeve speaks of—which set precedent for the reasonable time of editors of journals of literary scholarship up to this day—was two years.

What is instructive in this statement, however, is its acknowledgment that scholarly writing is done by people, prominent or not, who also have other things to do; that the very "prominence" of those who write is, in large part, determined by such publications. And that such publications, in fact, *justify* those institutions for which they speak by creating the possibility of conceiving of *knowledge* itself where before, in Gildersleeve's words, there existed simple a "struggle for notice amidst the miscellaneous matter of a review or the odds and ends of an educational magazine" (1880: 1). Thus, A. Marshall Elliott—the founding president of the Modern Language Association, whose name did not appear in the journal until the fifth volume—wrote in the first issue of *Modern Language Notes* in 1886 of the "pressing need" among the professors and teachers of modern languages "for some special organ of communication" to help describe this "branch of learning." In the same statement he situates scholarly publishing within the larger framework of the marketplace by promising the subscribers that "each number of the present volume will contain a least twenty-four columns of printed matter" (1886: title page). Similarly, half the articles in the first issue of his other journal, *PMLA*, articulate the *position* of modern languages as a branch of learning within the larger framework of the academy, what the opening article of its second volume calls the "Study of Modern Languages in our Higher Institutions."

Such editorial statements, including those of Conarroe and Cohen, are examples of what Foucault calls the "enunciatory modality" of discourse, and they *position* or *situate* the conduct of editors and journals within a particular order of things. Philip Lewis, in defining the "editorial-function" he assumed as editor of *Diacritics*, follows Foucault's description of the "author-function" as a particular enunciatory modality and describes three functions of an editor—and, implicitly, of publishing itself.

> The three kinds of ongoing pressure and constraint that I have just noted correspond, roughly, to three objects of the editor's awareness: the status of editing in academic work, the problems of management and marketing, and the irrecusability of ideological conflict. In the respective spheres of the university, the publishing business, and the scholarly field, the academic editor's activities and her/his views of the work to be done appear to be

> regulated by the operational principles of an institution, an economy, a profession. (1982: 229)

In their broadest terms, these categories follow Foucault's enunciatory, external, and internal constraints upon cognitive discourse. (That the institutional constraints of the university are, in fact, "enunciatory" is clear from Lewis's example of editing as positioned marginally, like linguistic deitics, within the institutional framework of compensation and reward [1982: 226].) In this editorial statement Lewis is doing what Miller claims each of the grounds of literary study always does, which is "to exercise sovereign control over the others, to make the others find their ground in *it*" (Miller 1985: 22). That is, Lewis is following Foucault in grounding the order of discourse in enunciatory discursive practice—practices of power. Conarroe also asserts the sovereignty of his editorial-discursive practice in grounding discourse in self-evident, socially perceivable truth, and Cohen does so as well, grounding discourse in his authoritative programmatic vision.

But a final form of editorial authority can be seen in the contrary to Foucault's enunciation. This is the *ironic* enunciation of a "fictional" or "marginal" editor, an enunciation which positions or grounds itself and simultaneously *shifts* that ground, linking it to others. Such "enunciation" is ironic insofar as it undermines or denies the *ontological* status of the editor altogether. Sören Kierkegaard is perhaps the best example of such editorial practice. In his pseudonymous works, he positions himself as the editor of fictional "authors." As such, the "authors" of these books are neither the transcendental authorities Conarroe describes nor Cohen's secondary figures; but neither do they embody or articulate instituted and institutional power. Rather, as the Judge says in *Either/Or,* they attempt to bring the reader "to the point where the choice between the evil and the good acquires significance" (1959: 172). In describing the purpose of his discourse, the Judge is describing Kierkegaard's ethical enterprise, his editorial conduct. By creating fictional authors, Kierkegaard defines his own larger career by offering the never-explicit relationships among his "authors" as positions to be chosen. The relationships among these authors are negative and polemical. But more important, the relationship between these authors and Kierkegaard's reader are even more negative—they are the contrary to Foucault's enunciatory power—precisely because they are *not* polemical, because they eschew their own explicit designs on the reader. Instead, they delimit, silently and ironically, the fourth ground Miller describes, a negativity "encountered in our relations to other people" (1985: 21).

In Kierkegaard we can find an element of obligation in editorial conduct emanating from an editorial position that is at once central and marginal, authorizing and authoritative and at the same time without authority and undermining the authority of hierarchy. Catharine Stimpson describes such an editorial position as an institutional practice in her retrospective description of the founding and editing of *Signs*. Stimpson begins her narrative with the assertion that "a journal editor must live in the future, with the next issue, the next volume, the next idea" (1982: 241). In an editorial in the first issue of *Signs* in 1975 she had said much the same thing: "a journal," the editors wrote, "is a series with a future" (1975: vii). By imagining editorial activity as *future* activity Stimpson allows for the truth and power of editing scholarly publications and, at the same time, underscores the element of obligation in it. What allows her to see—both retrospectively and prospectively—the activity of the editor as an activity that takes place in the future is precisely the feminist enterprise of creating a future: a new discipline, emerging "knowledge"—what Rachel Blau DuPlessis calls "shifting focus, bringing the world into different perspectives" because of the "situation of women," because of "our social situation, our relationship to power, our relationship to language" (1985: 285). This shifting focus, as we noted in Chapter 1, is central to the critique feminism brings to science as well. Moreover, this shifting focus is reinforced in the situation of editing a journal that is self-consciously feminist, scholarly, and, necessarily, political. "Editing *Signs*," Stimpson writes, "sets up special relationships to power. Our office has tried to embody certain feminist and egalitarian principles" (1982: 245). For this reason, even the retrospective narrative of her editorial practice is tentative: it includes a postscript that both narrates the very giving up of editorial authority and enacts the giving up of the authority of linear narrative. "Those of us who founded the journal," she notes, "decided that if we were feminists, we ought to share our positions with others and establish the practice of rotating the editorship every five years or so" (1982: 247).

Such a conception of editorial rotation—it exists, "fictionally," in Kierkegaard—embodies what Elaine Showalter calls "the modern female aesthetic . . . : fluid, nonlinear, decentralized, nonhierarchic, and many-voiced" (1985: 15). It is what DuPlessis calls "a both/and vision born of shifts, contraries, negations, contradictions. . . . Structurally, such a writing," she goes on, "might say different things, not settle on one, which is final. This is not a condition of 'not choosing,' since choice exists always in what to represent and in the rhythms of presentation. It is nonacademic" (1985: 276). This description of a feminist aesthetic also describes an editorial ethics in the curious and difficult balance between the whole and the parts, the authority of an editor, almost literally "authorizing" and creating authors, and the effacement of letting others speak for themselves and to others: an element of obligation. "In my essays' psychic and speculative search for contradictions, for wholeness," DuPlessis continues, "linear and constellated forms coexist. . . . The struggle with cultural hegemony, and the dilemmas of that struggle, are articulated in a voice that does not seek authority of tone or stasis of position but rather seeks to express the struggle in which it is immersed" (1985: 282–83).

In these four editorial positions we have presented four different understandings and conducts of the authority and authorization of cognition. Conarroe defines "authorization" as literally making a writer an author by publishing his or her work. Cohen defines it as the inclusion of particular authored manuscripts within his editorial vision, the authorization of work in terms of his stamp of approval. And Lewis subsumes the two by situating both within a particular framework of social, personal, and ultimately institutional discursive power, thus revealing editorial authorization to be institutionally defined functions—"editor-functions." Finally, Stimpson reframes these editor-functions, making them the object of an ethics. That is, she acknowledges, as does Lewis, that the role of editor—her very cognitive activity—is determined by certain institutional conventions and that the authority of the editor is a function of these conventions. But, unlike Lewis, she makes clear the frustrating conundrum faced by the editor who acknowledges the institutional source of editorial authority and the resulting ethical imbalance in the author-editor-reader relationship. Like Kierkegaard, her concern is to put into the hands of the reader the authority of judging *worth*. But the transference of such authority must be made within institutions, both cognitive and economic, that privilege the editor. Since it is not possible to eliminate the role of an editor and the institutional stage on which this role is acted out, Stimpson makes the role provisional—temporary—and with a kind of Kierkegaardian irony, neutralizes editorial authority.

Kierkegaard defines this role in the Socratic metaphor of "midwife," whose function is *to be ready* to assist at birth, to deliver into the hands of others, and whose activity is "maieutic" education. The editor, like the midwife-philosopher, follows the etymological sense of *educator*, letting child, pupil, author/reader *be* (Schleifer and Markley 1984: 17–20; see also Bové 1984). Like the midwife, the editor is both "there" and not there, marginal, at hand, in preparation for what could occur, what is only possible: local, emergent cognition. In this position, the fourth editorial role functions by means of the negative ontology of Miller's fourth ground.

In these terms, this editorial position, embodied in the feminist editor Stimpson describes, is ethical in Kristeva's post-Nietzschean definition insofar as the particular ends of editing are the goals of "shattering" the preexisting assumptions of readers, of offering them *writerly* pseudonymous texts that are, ironically, *not yet understood,* neither knowledge nor nonsense, neither the discourse of knowledge nor the discourse of power.

Here, indeed, in the future tense of this activity, the emergent *institutional* nature of cognition is made clear. The negative authorization of her practice situates the editor in a position to return discourse to her writers and readers: the editor, like Miller's "good reader" and even like Foucault himself, "unedits" the very principle of cognitive authority. Such "editorial" conduct questions, or "critiques" as Miller says, the grounds of editing and the grounds of cognitive knowledge. It forces us to recognize the existence of "editing" in all instituted authority, the element of power constituting knowledge, the fact that authors or visions do not speak for themselves. Such recognition is uncanny and strange, a kind of shock not of recognition but simply that something is not quite right, and it leads to an urge to domesticate it, an urge to the "choice" Kierkegaard attempts to bring his readers to, and also to the irony of undecidable, conflicting grounds for choice that, finally, Miller describes.

Still, irony, even irony acknowledged as a criterion and mode of understanding and conduct, cannot ground anything. Nothing rests in irony, least of all the cognitive activity of producing and disseminating "knowledge," which, like ethics itself—like the activity of the midwife—goes on and on. It is precisely the impossibility of restless irony serving as a ground of knowledge that constitutes the unsettled and unsettling conclusion Paul Bové reaches in his discussion of the ironic ethics—the "maieutics"—of Nietzsche's *Genealogy* (1986: 9–23). And it is precisely this impossibility that leads Bruno Latour to situate the signs of semiotics within the cultural network of scientific activity (1986: 26)—even while he acknowledges the convincing critique semiotics accomplishes on the conception of "scientific mind" that Whitehead and some strands of cognitive science use without analyzing. But if irony (and the circulation of signs in which irony traffics) can ground nothing, neither can ethics ground anything since the aim of ethics, as Dewey defines it, remains the detection of obligation and the examination of worth, responsive and responsible critique dealing with power and knowledge. Whether negative, responsive irony can ever accomplish the ends of the activity of cognition—whether it can ever be ethically responsible—is precisely the question left out of Miller's examination of the grounds of literary study. But a responsive and responsible critique, grounded in the social and personal knowledge of what is good and the ironic countermemory of the fact that such acknowledgment is an act of power and politics, might very well, among the others enunciated here, describe the ethics of instituted understanding.

Conclusion: Narrative Cognition

Paul Ricoeur's description of history in *Time and Narrative* reiterates the narrative process of cultural cognition we have figured in the person of the editor, and in so doing it underlines the "conclusion" to which the three parts of our study lead, what we have repeatedly suggested is the dynamic conception of the *emergence* not only of cognitive understanding but of the objects of cognition themselves. In the study of history, Ricoeur says, "historians do not proceed from the classificatory term toward the general law but from the classificatory term toward the explanation of differences" (1984: 124–25). This description is analogous to Greimas's semiotic square—whose analysis of narrative Ricoeur himself follows in fine detail (1985: 44–60)—the aim of which, as we have demonstrated, is the elaboration of difference within seeming semantic unities. But more important, Ricoeur's description of the work of historians also defines the activity—the goal—of an editor overseeing, maieutically, the emergence of "knowledge." When Ricoeur defines the historian's aim as the explanation of differences, he is articulating a goal that is analogous to Clifford Geertz's "thick description." "Explanation," Ricoeur argues, "cannot be converted into a prediction" (1984: 122); "the symmetry between explanation and prediction, characteristic of the nomological sciences, is broken at the very level of historical statements" (1984: 147). Instead of ending in prediction, in historical explanation "the divisibility of time ends where the most detailed analysis does" (1984: 123).

Such divisibility, like the semantic "explosion" effected by the semiotic square, "ends" in narrative. It does so because "hypotheses are not the goal of history, only landmarks for delineating a field of investigation, guides serving a mode of understanding which is fundamentally that of interpretative narrative, which is neither chronology nor 'science'" (1984: 156). Ricoeur is distinguishing between history and science, but we have argued that the concept of emergent cognition obviates or "neutralizes" this opposition. When understanding is conceived as "emergent"—whether it arises in opposition to received conceptions and paradigms, or in reimagining the modes by which it is taught and codified—it assumes the form of narrative understanding. A narrative, as we have seen, is more than events in serial order. Rather, narrative organizes "events" (which are themselves a function of narrative understanding) into an intelligible whole in which, as Ricoeur says, one can reasonably seek a controlling "thought." "Emplotment," he writes, "is the operation that draws a configuration out of a simple succession" (1984: 65). In this way the understandings of narrative refigure the temporal world in a mode of cultural construction that accomplishes "the effort of thinking which is at work in every narrative configuration" (1988: 3).

For this reason, the emergence of cognition is always interested, always goal oriented. Like the very conclusion we are proferring here, it is governed by the function of superimposing a temporal conclusion and an ideational (or ideological) "thought" or "point." "To follow a story," Ricoeur says,

> is to move forward in the midst of contingencies and peripeteia under the guidance of an expectation that finds its fulfilment in the "conclusion" of the story. This conclusion is not logically implied by some previous premises. It gives the story an "end point," which, in turn, furnishes the point of view from which the story can be perceived as forming a whole. To understand the story is to understand how and why the successive episodes led to this conclusion, which, far from being foreseeable, must finally be acceptable, as congruent with the episodes brought together by the story. (1984: 66–67)

Such a conclusion ends a narrative that is neither pure contiguity nor a logical implication. Conclusion and the narrative that arises with that conclusion offer what Ricoeur describes as a "third time" mediating between "lived time and cosmic time" (1988: 99), between the givenness of phenomenal experience and the data of the world. Such time is emergent, essentially incomplete, and finally cultural. Like Nietzsche's great genealogical project, it stands between "philology and physiology . . . [as] a theory of culture" (1988: 238).

The emergence of cognition, like the "institutions" of understanding we have traced in this chapter, is a form of evaluation and judgment implicit in the possibility Ricoeur describes of "'(as yet) untold' stories, stories that demand to be told, stories that offer anchorage points for narrative" (1984: 74). Ricoeur offers two examples of "(as yet) untold" stories. The first is the discourse of psychoanalysis—the stories and storytelling we described in Part II of this book. The second is more striking in terms of the narrative element we are emphasizing here in conclusion. It is the instance of a judge trying to understand a course of actions "by unraveling the tangle of plots the subject is caught up in" (1984: 74–75). Such plots, Ricoeur says, are entangled in the "prehistory" of the narrative being judged which "binds it to a larger whole and gives it a 'background.' This background is made up of the 'living imbrication' of every lived story with every other such story. Told stories therefore have to 'emerge' from this background" (1984: 75).

Narrative, that is, like cognition itself, begins with a horizon of expectation. In Greimas's case, it is expectation implied by the self-evident semantic unity of the first term of the semiotic square. In Darwin's case, it is expectation generated by a focus on accidental empirical "facts" and the simultaneous occultation of chance and accident implicit in his account of those "facts." In Freud's case, it is expectation inherent in the assumption of the explanatory power of narrative itself. Ricoeur calls this power the knowledge "that every narrative explains itself" because "to narrate what has happened is already to explain why it happened" (1984: 154). The horizons of expectation, more or less explicitly governing narrative, imply change and emergence simply by *responding* to the world with configurations and conclusions, putting the aporias of time and understanding to work to delineate the very "future" or "ends" from which narrative seems to operate. Such "conclusions" are always provisional. Greimas's logic is always provisional because his squares repeatedly generate new starting points for additional analysis. Accounts of adaptation, like more general accounts of the world, are always provisional because there is no end of discerning relationships within the environment. And the talking cure itself, as Freud says, is interminable because there is always something more to say. But horizons of expectation are also always provisional because in the negative discourse of the contrary-to-fact future orientation of expectation and emergence, they repeatedly create the possibility of more than a single account, of contest, trial, and judgment.

The work of narrative, in this way, is profoundly cultural. It is, as Clifford Geertz says of cultural studies, "intrinsically incomplete. And, . . . the more deeply it goes the less complete it is" (1973: 29). Similarly, Ricoeur says that "narrative discourse [is] intrinsically incomplete" (1984: 144). It is for this reason that the narrative understandings of cultural cognition always respond to and provoke other understandings. We have figured the subject of such understanding as editors, but Ricoeur, emphasizing the activity of narration implicit in this work of understanding, figures its subject as historians who, "like a judge, . . . are in a situation of contestation and of trial, [in which] their plea is never finished" (1984: 186). The situation of contestation and trial is the situation of *culture*—emergent, incomplete, and institutional. "The activities that define culture," Ricoeur writes, "are abstracted from particular societies and their modes are gathered together under a single classificatory concept by the definition that historians give to them, a definition that can vary widely from one author to another" (1984: 196). As such, the categories and "objects" of culture, including the cultural categories of cognition and understanding, are most fully grasped and comprehended by the work of narrative.

Such narratives, as we have said, traffic in "events" which are themselves grasped configurations, sites of contest, and results of judgment. The areas of cognitive activity we have examined in this chapter—the oppositional form of "knowledge," its pedagogical aim, and its institutional results—all describe forms of power configured with and contesting knowledge. That is, the irreducibility of power and knowledge to each other, like the irreducibility of special and general cases to each other in ethical considerations, marks the "areas" of cognitive activity we have examined as inevitably comprehended through narrative formats. Most important is the functioning of narrative in these "areas"—scientific structures of narrative in Part I, instances of particular narrative cognition in Part II, and the living imbrication of culture and cognition within the institutions of discourse we have examined here in Part III. From these disparate parts of our examination of cognition and the globally configured superimpositions of Anglo-American cognitive science, Continental semiotics, and Lacanian psychoanalysis, we are advancing the narrative "thought" of the constructed and contested activities of culture out of which emerge (among other things) knowledge and understanding.

Bibliography of Works Cited

Abrahams, J. P., and W. J. Hoyer, M. F. Elias, and B. Brandigan. 1975. "Gerontological Research in Psychology: Perspectives and Progress." *Journal of Gerontology* 30: 668–73.

Adams, Cynthia, Gisela Labouvie-Vief, Cathy J. Hobart, and Mary Dorosz. 1990. "Adult Age Group Differences in Story Recall Style." *Journal of Gerontology* 45: 17–27.

Aeschylus. 1959. *The Eumenides*. Trans. Richmond Lattimore. In *Aeschylus*. Chicago: University of Chicago Press.

Anton, John Peter. 1957. *Aristotle's Theory of Contrariety*. London: Routledge and Kegan Paul.

Apple, Michael W. 1982a. *Cultural and Economic Reproduction in Education*. London: Routledge and Kegan Paul.

———. 1982b. *Education and Power*. London: Routledge and Kegan Paul.

Apple, Michael W., and Lois Weis. 1983. *Ideology and Practice in Schooling*. Philadelphia: Temple University Press.

Aristotle. 1962. *On Interpretation*. Trans. Jean T. Oesterle. Milwaukee: Marquette University Press.

———. 1981. *Posterior Analytics*. Trans. Hippocrates G. Apostle. Grinnell, Iowa: Peripatetic Press.

———. 1989. *Poetics*. Trans. S. H. Butcher. In *Literary Criticism and Theory*, ed. Robert Con Davis and Laurie Finke, 60–83. New York: Longman.

Aronowitz, Stanley. 1988. *Science as Power: Discourse and Ideology in Modern Society*. Minneapolis: University of Minnesota Press.

Aronowitz, Stanley, and Henry A. Giroux. 1985. *Education under Siege*. South Hadley, Mass.: Bergin and Garvey.

Attig, M., and L. Hasher. 1980. "The Processing of Frequency of Occurrence Information by Adults." *Journal of Gerontology* 35: 66–69.

Attridge, Derek. 1988. *Peculiar Language: Literature as Difference from the Renaissance to James Joyce*. Ithaca: Cornell University Press.

Augustine. 1960. *The Confessions of St. Augustine*. Trans. John Ryan. New York: Penguin Books.

Austin, J. L. 1962. *How to Do Things with Words*. Cambridge, Mass.: Harvard University Press.

[Bakhtin, M. M.]/Volosinov, V. N. 1986a. *Marxism and the Philosophy of Language*. Trans. Ladislav Matejka and I. R. Titunik. Cambridge, Mass.: Harvard University Press.

———. 1986b. *Speech Genres and Other Late Essays*. Trans. Vern McGee. Austin: University of Texas Press.

Barnes, Barry. 1983. "On the Conventional Character of Knowledge and Cognition." In *Science Observed: Perspectives on the Social Study of Science*, ed. Karin Knorr-Cetina and Michael Mulkay, 19–51. London: Sage.

Barnet, Sylvan, et al. 1971. *Dictionary of Literary, Dramatic, and Cinematic Terms*. Boston: Little Brown.

Barthes, Roland. 1974. *S/Z*. Trans. Richard Miller. New York: Hill and Wang.

———. 1977. *Roland Barthes by Roland Barthes*. Trans. Richard Howard. New York: Hill and Wang.

Bartlett, F. C. 1932. *Remembering*. Cambridge: Cambridge University Press.

Baudrillard, Jean. 1975. *The Mirror of Production*. Trans. Mark Poster. St. Louis: Telos Press.

———. 1981. *For a Critique of the Political Economy of the Sign*. Trans. Charles Levin. St. Louis: Telos Press.

———. 1983. *Simulations*. Trans. Paul Foss, Paul Patton, Philip Beitchman. New York: Semiotext(e).

Beauvoir, Simone de. 1973. *The Coming of Age*. Trans. Patrick O'Brian. New York: Warner.

Bell, R. Q. 1971. "Stimulus Control of Parent or Caretaker Behavior of Offspring." *Developmental Psychology* 4: 63–72.

Bellemin-Noël, Jean. 1979. *Vers l'inconscient du texte*. Paris: Presses Universitaires de France.

Bengston, V. L. 1973. *The Social Psychology of Aging*. Indianapolis: Bobbs-Merrill.

Benjamin, Walter. 1969. *Illuminations*. Trans. Harry Zohn. New York: Schocken.

Benveniste, Emile. 1971. *Problems in General Linguistics*. Trans. Mary Elizabeth Meek. Coral Gables: University of Miami Press.

Bernstein, Richard J. 1983. *Beyond Objectivism and Relativism: Science, Hermeneutics, and Praxis*. Philadelphia: University of Pennsylvania Press.

Bickhard, M. H. 1979. "On Necessary and Specific Capabilities in Evolution and Development." *Human Development* 22: 217–24.

Birren, J. E. 1980. "Progress in Research on Aging in the Behavioral and Social Sciences." *Human Development* 23: 33–45.

Bleikasten, André. 1981. "Fathers in Faulkner." In Davis 1981: 115–46.

Blest, A. D. 1963. "Longevity, Palatability, and Natural Selection in New Species of New World Saturniid Moth." *Nature* 197: 1183–86.

Bloom, Allan. 1987. *The Closing of the American Mind*. New York: Simon and Schuster.

Bloomfield, Leonard. 1933. *Language*. New York: Holt, Rinehart, Winston.

Bloor, David. 1976. *Knowledge and Social Imagery*. London: Routledge and Kegan Paul.

Blythe, Ronald. 1979. *The View in Winter: Reflections on Old Age*. New York: Harcourt Brace Jovanovich.

Bono, James. 1990. "Science, Discourse, and Literature: The Role/Rule of Metaphor in Science." In *Literature and Science: Theory and Practice,* ed. Stuart Peterfreund, 59–89. Boston: Northeastern University Press.

Botwinick, Jack. 1961. "Husband and Father-in-law: A Reversible Figure." *American Journal of Psychology* 74: 312–13.

———. 1969. "Disinclination to Venture Response versus Cautiousness in Responding: Age Differences." *Journal of General Psychology* 115: 55–62.

———. 1978. *Aging and Behavior: A Comprehensive Integration of Research Findings*. 2d ed. New York: Springer.

Botwinick, Jack, and M. Storandt. 1974. *Memory, Related Functions, and Age*. Springfield: Thomas.

Bové, Paul. 1984. "The Penitentiary of Reflection: Sören Kierkegaard and Critical Activity." In *Kierkegaard and Literature,* ed. Ronald Schleifer and Robert Markley, 25–57. Norman: University of Oklahoma Press.

———. 1986. *Intellectuals in Power: A Genealogy of Critical Humanism*. New York: Columbia University Press.

Bowlby, John. 1958. "The Nature of the Child's Tie to His Mother." *International Journal of Psycho-Analysis* 39: 350–75.

Brantlinger, Patrick. 1990. *Crusoe's Footprints: Cultural Studies in Britain and America*. New York: Routledge.

Brent, S. B. 1978. "Individual Specialization, Collective Adaptation, and Rate of Environmental Change." *Human Development* 21: 21–33.

Brinley, J. F. 1965. "Cognitive Sets and Accuracy of Performance in the Elderly." In *Behavior, Aging, and the Nervous System,* ed. A. T. Welford and J. E. Birren, 114–49. Springfield: Thomas.

Brooks, Peter. 1976. *The Melodramatic Imagination*. New Haven: Yale University Press.

———. 1979. "Fiction of the Wolfman: Freud and Narrative Understanding." *Diacritics* 9: 72–81.

Brubaker, T. H., and E. A. Powers. 1976. "The Stereotype of 'Old.'" *Journal of Gerontology* 31: 441–47.

Bruner, Jerome. 1986. *Actual Minds, Possible Worlds*. Cambridge, Mass.: Harvard University Press.

———. 1990. "Culture and Human Development: A New Look." *Human Development* 33: 344–55.

Burke, Kenneth. 1969. "Four Master Tropes." In *A Grammar of Motives,* 503–17. Berkeley and Los Angeles: University of California Press.

Butler, R. N. 1963. "The Life Review: An Interpretation of Reminiscence in the Aged." *Psychiatry* 26: 65–76.

Ceci, Stephen J., and Steven W. Cornelius. 1990. "Commentary." *Human Development* 33: 198–201.

Charlip, W. S. 1968. "The Aging Female Voice: Selected Fundamental Frequency Characteristics and Listener Judgment." Ph.D. diss., Purdue University, Lafayette, Ind.

Chekhov, Anton. 1979. "The Lady with the Dog." Trans. Ivy Litvinov. In *Anton Chekhov's Short Stories*, ed. Ralph Matlaw, 221–34. New York: Norton.

Christian, R. R., and G. T. Baker. 1979. "Influence of Post-reproductive Cohorts in the Selection for Increased Longevity." Paper presented at the Annual Meeting of the Gerontological Society, Washington, D.C.

Clausen, J. A. 1968. "The Life Course of Individuals." In *Aging and Society:* vol. 3, *A Sociology of Age Stratification*, ed. M. W. Riley and Anne Foner. New York: Russell Sage.

Cockburn, Janet, and Philip Smith. 1991. "The Relative Influence of Intelligence and Age on Everyday Memory." *Journal of Gerontology* 46: 31–36.

Cohen, D. 1980. "Donald Hebb: An Inside Look at Aging." *APA Monitor*, February: 4–5.

Cohen, Ralph. 1982. "On a Decade of *New Literary History*." In *The Horizon of Literature*, ed. Paul Hernadi, 249–59. Lincoln: University of Nebraska Press.

Conarroe, Joel. 1982. "A Galaxy of Editors." In *The Horizon of Literature*, ed. Paul Hernadi, 213–19. Lincoln: University of Nebraska Press.

Conrad, Joseph. 1971. *Heart of Darkness*. Ed. Robert Kimbrough. New York: Norton.

Cooper, Patricia. 1990. "Discourse Production and Normal Aging: Performance on Oral Picture Description Tasks." *Journal of Gerontology* 45: 210–14.

Corso, J. F. 1977. "Auditory Perception and Communication." In *Handbook of the Psychology of Aging*, ed. J. E. Birren and K. W. Schaie, 535–53. New York: Van Nostrand Reinhold.

Coward, Rosalind, and John Ellis. 1977. *Language and Materialism: Developments in Semiology and the Theory of the Subject*. London: Routledge and Kegan Paul.

Craik, F. I. M., and R. S. Lockhart. 1972. "Levels of Processing: A Framework for Memory Research." *Journal of Verbal Learning and Verbal Behavior* 11: 671–84.

Culler, Jonathan. 1981. *The Pursuit of Signs: Semiotics, Literature, Deconstruction*. Ithaca: Cornell University Press.

———. 1982. *On Deconstruction: Theory and Criticism after Structuralism*. Ithaca: Cornell University Press.

Darwin, Charles. 1958. *The Origin of Species*. New York: New American Library.

Davidson, Donald. 1974. "On the Very Idea of a Conceptual Scheme." *Proceedings of the American Philosophical Association* 47: 5–20.

———. 1978. "What Metaphors Mean." *Critical Inquiry* 5: 31–47.

———. 1980. *Essays on Actions and Events*. Clarendon: Oxford University Press.

Davis, Robert Con, ed. 1981. *The Fictional Father: Lacanian Readings of the Text*. Amherst: University of Massachusetts Press.

———, ed. 1984. *Lacan and Narration: The Psychoanalytic Difference in Narrative Theory*. Baltimore: Johns Hopkins University Press.

Davis, Robert Con, and Ronald Schleifer. 1991. *Criticism and Culture: The Role of Critique in Modern Literary Theory*. London: Longman.

Dawkins, R. 1976. *The Selfish Gene*. New York: Oxford University Press.

Deleuze, Gilles. 1984. *Kant's Critical Philosophy: The Doctrine of the Faculties*. Trans. Hugh Tomlinson and Barbara Habberjam. Minneapolis: University of Minnesota Press.

De Man, Paul. 1969. "The Rhetoric of Temporality." In *Interpretation: Theory and Practice*, ed. Charles Singleton, 173–210. Baltimore: Johns Hopkins University Press.

———. 1979. *Allegories of Reading: Figural Language in Rousseau, Nietzsche, Rilke, and Proust*. New Haven: Yale University Press.

———. 1984. *The Rhetoric of Romanticism*. New York: Columbia University Press.

———. 1986. *The Resistance to Theory*. Minneapolis: University of Minnesota Press.

Derrida, Jacques. 1976. *Of Grammatology*. Trans. Gayatri Spivak. Baltimore: Johns Hopkins University Press.

———. 1978. *Writing and Difference*. Trans. Alan Bass. Chicago: University of Chicago Press,

———. 1979. *Spurs: Nietzsche's Styles*. Trans. Barbara Harlow. Chicago: University of Chicago Press.

———. 1981. *Positions*. Trans. Alan Bass. Chicago: University of Chicago Press.

———. 1982. *Margins of Philosophy*. Trans. Alan Bass. Chicago: University of Chicago Press.

———. 1987. *The Post Card: From Socrates to Freud and Beyond*. Trans. Alan Bass. Chicago: University of Chicago Press.

Dewey, John. 1969. *Outlines of a Critical Theory of Ethics*. In *Early Works*, volume 3, 1889–1892, ed. Jo Ann Boydston, 239–388. Carbondale: Southern Illinois University Press.

Dobzhansky, Theodosius. 1972. "Genetics and the Diversity of Behavior." *American Psychologist* 27: 523–30.

Dolen, L. S., and D. J. Bearison. 1982. "Social Interaction and Social Cognition in Aging: A Contextual Analysis." *Human Development* 25: 430–42.

Donne, John. 1967. *Poetical Works*. Ed. Herbert Grierson. Oxford: Clarendon Press.

Drieman, G. H. J. 1962a. "Differences between Written and Spoken Language: Qualitative Approach." *Acta psychologia* 10: 78–97.

———. 1962b. "Differences between Written and Spoken Language: Quantitative Approach." *Acta psychologia* 10: 36–57.

D'Souza, Dinesh. 1991. *Illiberal Education*. New York: Free Press.

duBois, Page. 1982. *Centaurs and Amazons: Women and the Pre-History of the Great Chain of Being*. Ann Arbor: University of Michigan Press.

DuPlessis, Rachel Blau. 1985. "For the Etruscans." In *The New Feminist Criticism*, ed. Elaine Showalter, 271–91. New York: Pantheon Books.

Durand, Régis. 1981. "'The Captive King': The Absent Father in Melville's Texts." In Davis 1981: 42–79.

Eagleton, Terry. 1990. *The Ideology of the Aesthetic*. Cambridge, Mass.: Basil Blackwell.

Eco, Umberto. 1976. *A Theory of Semiotics*. Bloomington: Indiana University Press.

Elders, Fons, ed. 1974. Interview: "Noam Chomsky and Michel Foucault." In *Reflexive Water: The Basic Concerns of Mankind.* London: Souvenir Press.

Eliot, George. 1961. *The Mill on the Floss.* Boston: Houghton Mifflin.

Eliot, T. S. 1975. *Selected Prose.* Ed. Frank Kermode. New York: Harcourt Brace.

Elliott, A. Marshall. 1886. "Editorial Note." *Modern Language Notes* 1: title page.

Erikson, Erik. 1962. *Childhood and Society.* New York: Norton.

Eysenck, M. W. 1974. "Age Differences in Incidental Learning." *Developmental Psychology* 10: 936–41.

Felman, Shoshana. 1977. "To Open the Question." *Yale French Studies* 55/56: 5–10.

____. 1983. *The Literary Speech Act: Don Juan with J. L. Austin, or Seduction in Two Languages.* Trans. Catherine Porter. Ithaca: Cornell University Press.

Feroleto, J. A., and B. R. Gounard. 1975. "The Effects of Subjects' Age and Expectations Regarding an Interviewer on Personal Space." *Experimental Aging Research* 1: 57–61.

Fineman, Joel. 1981. "The Structure of Allegorical Desire." In *Allegory and Representation,* ed. Stephen Greenblatt, 26–60. Baltimore: Johns Hopkins University Press.

Fish, Stanley. 1992. "There's No Such Thing as Free Speech and It's a Good Thing, Too." In *Debating P.C.,* ed. Paul Berman, 231–45. New York: Dell.

Foley, Barbara. 1985. "The Politics of Deconstruction." In *Rhetoric and Form: Deconstruction at Yale,* ed. Robert Con Davis and Ronald Schleifer, 113–34. Norman: University of Oklahoma Press.

Foucault, Michel. 1972a. *The Archaeology of Knowledge.* Trans. A. M. Sheridan Smith. New York: Harper.

____. 1972b. *The Discourse on Language.* Trans. Rupert Swyer. In *The Archaeology of Knowledge,* 215–37. New York: Harper.

____. 1977a. *Discipline and Punish.* Trans. Alan Sheridan. New York: Pantheon.

____. 1977b. *Language, Counter-Memory, Practice: Selected Essays and Interviews.* Trans. Donald Bouchard and Sherry Simon. Ithaca: Cornell University Press.

Fowler, Alastair. 1982. *Kinds of Literature: An Introduction to the Theory of Genres and Modes.* Cambridge, Mass.: Harvard University Press.

Fozard, J. L., E. Wolf, B. Bell, R. A. McFarland, and S. Podolsky. 1977. "Visual Perception and Communication." In *Handbook of the Psychology of Aging,* ed. J. E. Birren and K. W. Schaie, 497–534. New York: Van Nostrand.

Freire, Paulo. 1982. *Pedagogy of the Oppressed.* Trans. Myra Berman Ramos. New York: Continuum.

____. 1985. *The Politics of Education: Culture, Power, and Liberation.* Trans. Donaldo Macedo. South Hadley, Mass.: Bergin and Garvey.

Freud, Sigmund. 1963. "Neurosis and Psychosis." Trans. Joan Riviere. In *General Psychoanalytical Theory,* ed. Philip Rieff, 185–89. New York: Collier Books.

____. 1965. *The Interpretation of Dreams.* Trans. James Strachey. New York: Avon.

Frye, Northrop. 1957. *Anatomy of Criticism.* Princeton: Princeton University Press.

Gadow, Sally. 1983. "Frailty and Strength: The Dialectic of Aging." *Gerontologist* 23: 144–47.

Galan, F. W. 1985. *Historic Structures: The Prague School Project, 1928–1946*. Austin: University of Texas Press.

Gallagher, Catherine. 1985. "Politics, the Profession, and the Critic." *Diacritics* 15, 2: 37–43.

Gallop, Jane. 1982. "The Immoral Teachers." *Yale French Studies* 63: 117–128.

Gardner, Howard. 1985. *The Mind's New Science: A History of the Cognitive Revolution*. New York: Basic.

Garfinkel, Harold. 1967. *Studies in Ethnomethodology*. Englewood-Cliffs: Prentice-Hall.

Geertz, Clifford. 1973. *The Interpretation of Cultures*. New York: Harper Torchbooks.

Gibson, Eleanor J., and Elizabeth S. Spelke. 1983. "The Development of Perception." In *Handbook of Child Psychology*, 4th edition, ed. Paul H. Mussen, volume 3, *Cognitive Development*, ed. John H. Flavell and Ellen M. Markman, 1–76. New York: John Wiley.

Gildea, P., and S. Glucksberg. 1983. "On Understanding Metaphor: The Role of Context." *Journal of Verbal Learning and Verbal Behavior* 22: 577–90.

Gildersleeve, B. L. 1880. "Editorial Note." *American Journal of Philology* 1: 1–2.

Giroux, Henry, David Shumway, Paul Smith, and James Sosnoski. 1984. "The Need for Cultural Studies: Resisting Intellectual and Oppositional Public Spheres." *Dalhousie Review* 64: 472–86.

Glucksberg, S., P. Gildea, and H. B. Bookin. 1982. "On Understanding Nonliteral Speech: Can People Ignore Metaphors?" *Journal of Verbal Learning and Verbal Behavior* 21: 85–98.

Goody, Jack. 1976. "Aging in Nonindustrial Societies." In *Handbook of Aging and the Social Sciences*, ed. R. H. Binstock and Ethel Shanas, 117–29. New York: Van Nostrand.

Gould, Stephen Jay. 1977. *Ever Since Darwin: Reflections in Natural History*. New York: Norton.

———. 1981. *The Mismeasure of Man*. New York: Norton.

Graff, Gerald. 1987. *Professing Literature: An Institutional History*. Chicago: University of Chicago Press.

Gramsci, Antonio. 1985. *Selections from Cultural Writings*. Ed. David Forgacs, Geoffrey Nowell-Smith. Trans. William Boelhowever. Cambridge, Mass.: Harvard University Press.

Gregory, Monica, and Nancy Mergler. 1990. "Metaphor Comprehension: In Search of Literal Truth, Possible Sense, and Metaphoricity." *Metaphor and Symbolic Activity* 5: 151–73.

Greimas, A. J. 1970. *Du Sens*. Paris: Seuil.

———. 1974. "Interview" in *Discussing Language*, ed. Herman Parret, 55–71. The Hague: Mouton. 55–71.

———. 1983a. *Du sens II*. Paris: Seuil.

———. 1983b. *Structural Semantics: An Attempt at a Method*. Trans. Daniele MacDowell, Ronald Schleifer, and Alan Velie. Lincoln: University of Nebraska Press.

———. 1987. *On Meaning: Selected Writings in Semiotic Theory.* Trans. Paul Perron and Frank Collins. Minneapolis: University of Minnesota Press.

———. 1988. *Maupassant: The Semiotics of Text.* Trans. Paul Perron. Amsterdam and Philadelphia: John Benjamins.

———. 1989a. "Description and Narrativity: 'The Piece of String.'" Trans. Paul Perron and Frank Collins. *New Literary History* 20: 615–26.

———. 1989b. "On Meaning." Trans. Paul Perron and Frank Collins. *New Literary History* 20: 539–50.

Greimas, A. J., and J. Courtés. 1982. *Semiotics and Language: An Analytical Dictionary.* Trans. Larry Crist, Daniel Patte, et al. Bloomington: Indiana University Press.

Greimas, A. J., and Eric Landowski. 1976. "Analysis sémiotique d'un discours juridique." In *Sémiotique et sciences sociales,* 79–128. Paris: Seuil.

Gubrium, J. F. 1976. *Time, Roles, and Self in Old Age.* New York: Human Sciences Press.

Guetti, Barbara. 1982. "The Old Regime and the Feminist Revolution: Laclos' 'De l'Education des Femmes.'" *Yale French Studies* 63: 139–62.

Haecan, H., and M. L. Albert. 1978. *Human Neuropsychology.* New York: Wiley.

Hall, Stuart. 1980. "Cultural Studies: Two Paradigms." *Media, Culture and Society* 2: 57–72.

Haraway, Donna. 1991. "Situated Knowledges: The Science Question in Feminism and the Privilege of Partial Perspective." In *Simians, Cyborgs, and Women,* 183–201. New York: Routledge.

Harding, Sandra. 1986. "The Instability of the Analytical Categories of Feminist Theory." *Signs* 11: 645–64.

Harris, Zellig. 1951. *Methods in Structural Linguistics.* Chicago: University of Chicago Press.

Hartman, Geoffrey. 1975. *The Fate of Reading.* Chicago: University of Chicago Press.

———. 1980. *Criticism in the Wilderness.* New Haven: Yale University Press.

Hasher, Lynn, and R. T. Zacks. 1979. "Automatic and Effortful Processing in Memory." *Journal of Experimental Psychology: General* 108: 356–88.

Havelock, Eric. 1963. *Preface to Plato.* Cambridge, Mass.: Harvard University Press.

Havighurst, R. J. 1968. "Personality and Patterns of Aging." *Gerontologist* 8: 20–23.

Hawkesworth, Mary. 1989. "Knowers, Knowing, Known: Feminist Theory and Claims of Truth." *Signs* 14: 533–57.

Hayles, N. Katherine. 1990. *Chaos Bound: Orderly Disorder in Contemporary Literature and Science.* Ithaca: Cornell University Press.

———. 1991. "Constrained Constructivism: Locating Scientific Inquiry in the Theater of Representation." In *Interphysics: Postdisciplinary Approaches to Literature and Science,* ed. Robert Markley. *New Orleans Review* 18, 1: 76–85.

Heidegger, Martin. 1972. *On Time and Being.* Trans. Joan Stambaugh. New York: Harper and Row.

Hirst, William, and David Kalmar. 1987. "Characterizing Attentional Resources." *Journal of Experimental Psychology* 116: 68–81.

Hjelmslev, Louis. 1961. *Prolegomena for a Theory of Language*. Trans. Francis Whitfield. Madison: University of Wisconsin Press.

Howes, Janice, and Albert Katz. 1988. "Assessing Remote Memory with an Improved Public Events Questionnaire." *Psychology and Aging* 3: 142–50.

Hultsch, David, Michael Masson, and Brent Small. 1991. "Adult Age Differences in Direct and Indirect Tasks of Memory." *Journal of Gerontology* 46: 22–30.

Illich, Ivan. 1970. *Deschooling Society*. New York: Harper and Row.

Jaeger, Werner. 1974. *Paideia, Volume 1*. Trans. Gilbert Highet. New York: Oxford University Press.

Jakobson, Roman. 1962. "The Concept of the Sound Law and the Teleological Criterion." In *Selected Writings: Volume 1, Philological Studies*, 1–2. The Hague: Mouton.

Jakobson, Roman, and Morris Halle. 1971. *Fundamentals of Language*. The Hague: Mouton.

Jameson, Fredric. 1972. *The Prison-House of Language*. Princeton: Princeton University Press.

———. 1977. "Imaginary and Symbolic in Lacan: Marxism, Psychoanalytic Criticism, and the Problem of the Subject." *Yale French Studies* 55/56: 338–95.

———. 1981. *The Political Unconscious: Narrative as a Socially Symbolic Act*. Ithaca: Cornell University Press.

———. 1987. "Foreword" to A. J. Greimas 1987: vi–xxii.

Johnson, Barbara. 1977. "The Frame of Reference." *Yale French Studies* 55/56: 457–505.

———. 1982. "Teaching Ignorance: *L'Ecole des Femmes*." *Yale French Studies* 63: 165–82.

Joyce, James. 1967. *Dubliners*. New York: Penguin.

Karl, C. S., E. S. Metress, and J. F. Metress. 1978. *Aging and Health: Biologic and Social Perspectives*. Menlo Park: Addison-Wesley.

Keller, Evelyn Fox. 1982. "Feminism and Science." In *Feminist Theory: A Critique of Ideology*, ed. Michelle Rosaldo and Barbara Gelpi, 113–26. Chicago: University of Chicago Press. 113–26.

———. 1985. *Reflections on Gender and Science*. New Haven: Yale University Press.

Kellner, Hans. 1981. "The Inflatable Trope as Narrative Theory: Structure or Allegory." *Diacritics* 11, 1: 14–28.

Kenner, Hugh. 1978. *Joyce's Voices*. Berkeley and Los Angeles: University of California Press.

Kermode, Frank. 1969. *The Sense of an Ending*. New York: Oxford University Press.

———. 1979. *The Genesis of Secrecy: On the Interpretation of Narrative*. Cambridge, Mass.: Harvard University Press.

Kern, Stephen. 1983. *The Culture of Time and Space: 1880–1918*. Cambridge: Harvard University Press.

Kierkegaard, Sören. 1959. *Either/Or*. Volume 2. Trans. David Swenson and Lillian Swenson. New York: Anchor.

Kirkwood, T. B. L., and R. Holliday. 1979. "The Evolution of Aging and Longevity." *Proceedings of the Royal Society, London* 205: 531–46.

Kristeva, Julia. 1980. *Desire in Language: A Semiotic Approach to Literature and Art*. Trans. Thomas Gora, Alice Jardine, Leon Roudiez. New York: Columbia University Press.

Kuczaj, S. A., B. Harbaugh, and R. Boston. 1979. "What Children Think about the Speaking Capabilities of Other Persons and Things." Paper presented at the Annual Meeting of the Psychonomic Society, Phoenix.

Kuhn, Thomas S. 1970. *The Structure of Scientific Revolutions*. Chicago: University of Chicago Press.

———. 1977. *The Essential Tension: Selected Studies in Scientific Tradition and Change*. Chicago: University of Chicago Press.

Labouvie-Vief, Gisela. 1977. "Adult Cognitive Development: In Search of Alternative Interpretations." *Merrill-Palmer Quarterly* 23: 227–63.

———. 1980. "Beyond Formal Operations: Uses and Limits of Pure Logic in Life-span Development." *Human Development* 23: 141–61.

———. 1982. "Growth and Aging in Life-span Perspective." *Human Development* 25: 65–69.

Labouvie-Vief, Gisela, and J. N. Gonda. 1976. "Cognitive Strategy Training and Intellectual Performance in the Elderly." *Journal of Gerontology* 31: 327–32.

Lacan, Jacques. 1972. "Seminar on 'The Purloined Letter.'" Trans. Jeffrey Mehlman. *Yale French Studies* 48: 38–72.

———. 1977a. "Desire and the Interpretation of Desire in *Hamlet*." Trans. James Hulbert. *Yale French Studies* 55/56: 11–52.

———. 1977b. *Ecrits: A Selection*. Trans. Alan Sheridan. New York: Norton.

———. 1978. *The Four Fundamental Concepts of Psycho-Analysis*. Trans. Alan Sheridan. New York: Norton.

———. 1991. *The Seminar of Jacques Lacan: Book II. The Ego in Freud's Theory and in the Technique of Psychoanalysis, 1954–1955*. Trans. Sylvana Tomaselli. New York: Norton.

Lass, N. J., P. J. Barry, R. A. Reed, J. M. Walsh, and T. A. Amuso. 1979. "The Effect of Temporal Speech Alterations on Speaker Height and Weight Identifications." *Language and Speech* 22: 163–71.

Lass, N. J., and M. Davis. 1976. "An Investigation of Speaker Height and Weight Identification." *Journal of the Acoustical Society of America* 59: 700–703.

Latour, Bruno. 1986. "Visualization and Cognition: Thinking with Eyes and Hands." *Knowledge and Society: Studies in the Sociology of Culture Past and Present* 6: 1–40.

———. 1987. *Science in Action: How to Follow Scientists and Engineers through Society*. Cambridge, Mass.: Harvard University Press.

———. 1988. "A Relativistic Account of Einstein's Relativity." *Social Studies of Science* 18: 3–44.

Latour, Bruno, and Steve Woolgar. 1986. *Laboratory Life: The Social Construction of Scientific Fact*. Princeton: Princeton University Press.

Layton, B. 1975. "Perceptual Noise and Aging." *Psychological Bulletin* 82: 875–83.

Leach, Edmund. 1970. *Claude Lévi-Strauss*. New York: Viking.

Lentricchia, Frank. 1983. *Criticism and Social Change*. Chicago: University of Chicago Press.

Levinson, Daniel, with C. Darrow, E. Klein, M. Levinson, and B. McKee. 1978. *The Seasons of a Man's Life*. New York: Knopf.

Lévi-Strauss, Claude. 1975. *The Raw and the Cooked*. Trans. John Weightman and Doreen Weightman. New York: Harper and Row.

——. 1984. "Structure and Form: Reflections on a Work by Vladimir Propp." Trans. Monique Layton, rev. Anatoly Liberman. In Vladimir Propp, *Theory and History of Folklore*, 167–89. Minneapolis: University of Minnesota Press.

Lewis, Philip. 1982. "Notes on the Editor-Function." In *The Horizon of Literature*, ed. Paul Hernadi, 221–40. Lincoln: University of Nebraska Press.

Liberman, Anatoly. 1984. "Introduction." In Vladimir Propp, *Theory and History of Folklore*, ix–lxxxi. Minneapolis: University of Minnesota Press.

Liss, Julie, Gary Weismer, and John Rosenbeck. 1990. "Selected Acoustic Characteristics of Speech Production in Very Old Males." *Journal of Gerontology* 45: 35–45.

Livingstone, David W. 1983. *Class Ideologies and Educational Futures*. Sussex: Falmer Press.

Loewen, E. R., R. J. Shaw, and F. I. M. Craik. 1990. "Age Differences in Components of Metamemory." *Experimental Aging Research* 16: 43–48.

MacCabe, Colin. 1975. *James Joyce and the Revolution of the Word*. London: New Left Books.

Mac Cormac, Earl. 1985. *A Cognitive Theory of Metaphor*. Cambridge, Mass.: MIT Press.

MacIntyre, Alisdair. 1977. "Epistemological Crises, Dramatic Narrative, and the Philosophy of Science." *The Monist* 60: 451–72.

McTavish, D. G. 1971. "Perceptions of Old People: A Review of Research Methodologies and Findings." *Gerontologist* 11: 90–101.

Mayer, P. J. 1979. "The Evolution of Human Longevity: Toward a Biocultural Theory." Paper presented at the Annual Meeting of the Gerontological Society, Washington, D.C.

Medawar, P. B. 1952. *An Unsolved Problem in Biology*. London: Kewls.

Mehlman, Jeffrey. 1981. "Trimethylamin: Notes on Freud's Specimen Dream." In *Untying the Text: A Post-Structuralist Reader*, ed. Robert Young, 177–88. London: Routledge and Kegan Paul.

Melville, Herman. 1967. *The Confidence-Man*. Indianapolis: Bobbs-Merrill.

Mergler, N. L., J. B. Dusek, and W. J. Hoyer. 1977. "Central/Incidental Recall and Selective Attention in Young and Elderly Adults." *Experimental Aging Research* 3: 49–60.

Mergler, N. L., Marion Faust, and M. D. Goldstein. 1985. "Storytelling as an Age-Dependent Skill: Oral Recall of Orally Presented Stories." *International Journal of Aging and Human Development* 20: 205–28.

Mergler, Nancy, and Ronald Schleifer. 1983. "Generational Myths: The Adult Child Caught in the Middle." *Centerboard: The Journal of the Southwest Center for Human Relations Studies* 1:62–68.

Merod, Jim. 1987. *The Political Responsibility of the Critic*. Ithaca: Cornell University Press.

Miller, George. 1956. "The Magical Number Seven, Plus or Minus Two: Some Limits on Our Capacity for Processing Information." *Psychological Review* 63: 81–97.

Miller, J. Hillis. 1976. "Stevens' Rock and Criticism as Cure." *Georgia Review* 30: 5–31, 330–48.

———. 1985. "The Search for Grounds in Literary Study." In *Rhetoric and Form: Deconstruction at Yale*, ed. Robert Con Davis and Ronald Schleifer, 19–36. Norman: University of Oklahoma Press.

Mink, Louis O. 1970. "History and Fiction as Modes of Comprehension." *New Literary History* 1: 541–58.

Mitchell, Juliet. 1974. *Psycho-Analysis and Feminism*. New York: Pantheon.

Myers, Greg. 1990. *Writing Biology: Texts in the Social Construction of Scientific Knowledge*. Madison: University of Wisconsin Press.

Mysak, E. D. 1959. "Pitch and Duration Characteristics of Older Males." *Journal of Speech and Hearing Research* 2: 46–54.

Navon, David, and Daniel Gopher. 1979. "On the Economy of the Human Processing System." *Psychological Review* 86: 214–55.

Nebes, R. D. 1978. "Vocal Versus Manual Response as a Determinant of Age Difference in Simple Reaction Time." *Journal of Gerontology* 33: 884–89.

Neisser, Ulric. 1982. *Memory Observed: Remembering in Natural Contexts*. San Francisco: W. H. Freeman.

———, ed. 1987. *Concepts and Conceptual Development: Ecological and Intellectual Factors in Categorization*. Cambridge: Cambridge University Press.

Neugarten, B. L., and G. O. Hagestad. 1976. "Age and the Life Course." In *Handbook of Aging and the Social Sciences*, ed. R. H. Binstock and Ethel Shanas, 35–55. New York: Van Nostrand.

Nietzsche, Friedrich. 1957. *The Use and Abuse of History*. Trans. Adrian Collins. Indianapolis: Bobbs-Merrill.

———. 1967. *The Genealogy of Morals*. Trans. Walter Kaufmann. New York: Vintage.

Norris, Christopher. 1988. *Paul de Man: Deconstruction and the Critique of Aesthetic Ideology*. New York: Routledge.

Nussbaum, Jon F., Teresa Thompson, and James Robinson. 1989. *Communication and Aging*. New York: Harper and Row.

Obler, L. K. 1980. "Narrative Discourse Style in the Elderly." In *Language and Communication in the Elderly*, ed. L. K. Obler and M. L. Albert, 75–90. New York: Lexington.

Ong, Walter. 1982. *Orality and Literacy: The Technologizing of the Word*. New York: Methuen.

Ortony, Anthony. 1979. "Beyond Literal Similarity." *Psychological Review* 86: 161–80.

Ortony, Anthony, D. L. Schallert, R. E. Reynolds, and S. J. Antos. 1978. "Interpreting Metaphors and Idioms: Some Effects of Context on Comprehension." *Journal of Verbal Learning and Verbal Behavior* 17: 465–77.

Ortony, Anthony, and R. J. Vondruska, M. A. Foss, and L. E. Jone. 1985. "Salience, Similes, and the Asymmetry." *Journal of Memory and Language* 24: 569–94.

Parret, Herman. 1983. *Semiotics and Pragmatics: An Evaluative Comparison of Conceptual Frameworks*. Amsterdam: John Benjamins.

Pearce, Roy Harvey. 1965. "Wallace Stevens: The Last Lesson of the Master." In *The Act of the Mind: Essays on the Poetry of Wallace Stevens*, ed. Roy Harvey Pearce and J. Hillis Miller, 121–42. Baltimore: Johns Hopkins University Press.

Percy, Walker. 1979. *The Message in the Bottle*. New York: Farrar, Straus, and Giroux.

Perlmutter, Marion. 1978. "What Is Memory Aging the Aging Of?" *Developmental Psychology* 14:330–45.

Perlmutter, Marion, Michael Kaplan, and Linda Nyquist. 1990. "Development of Adaptive Competence in Adulthood." *Human Development* 33:185–97.

Peterfreund, Stuart. 1990a. "Blake, Priestley, and the 'Gnostic Moment.'" In *Literature and Science: Theory and Practice*, ed. Stuart Peterfreund, 139–67. Boston: Northeastern University Press.

———. 1990b. "Wordsworth and Newtonian Time." *Genre* 23:279–96.

———. 1991. "Power Tropes." In *Interphysics: Postdisciplinary Approaches to Literature and Science*, ed. Robert Markley. *New Orleans Review* 18, 1: 27–36.

Piaget, Jean. 1970. *Structuralism*. Trans. Chaninah Maschler. New York: Basic.

Propp, Vladimir. 1968. *Morphology of the Folktale*. Trans. Laurence Scott, rev. Louis Wagner. Austin: University of Texas Press.

Quine, Willard Van Orman. 1961. *From a Logical Point of View*. New York: Harper Torchbooks.

Ricoeur, Paul. 1980. "Narrative Time." *Critical Inquiry* 7: 169–90.

———. 1984. *Time and Narrative*. Volume 1. Trans. Kathleen McLaughlin and David Pellauer. Chicago: University of Chicago Press.

———. 1985. *Time and Narrative*. Volume 2. Trans. Kathleen McLaughlin and David Pellauer. Chicago: University of Chicago Press.

———. 1988. *Time and Narrative*. Volume 3. Trans. Kathleen Blamey and David Pellauer. Chicago: University of Chicago Press.

Riegel, K. F., and R. M. Riegel. 1960. "A Study of Changes of Attitudes and Interests during Later Years of Life." *Vita Humanitas* 3: 177–206.

Riley, M. W. 1978. "Aging, Social Change, and the Power of Ideas." *Daedalus* 107: 39–52.

Romaniuk, M., J. G. Romaniuk, and Eric Labouvie. 1978. "The Content of the Elderly's Reminiscence: A Descriptive Analysis." Paper presented at the Annual Meeting of the Gerontological Society, Dallas.

Rorty, Richard. 1979. *Philosophy and the Mirror of Nature*. Princeton: Princeton University Press.

———. 1982. *Consequences of Pragmatism*. Minneapolis: University of Minnesota Press.

Rose, Jacqueline. 1982. "Introduction—II." In *Jacques Lacan and the école freudienne, Feminine Sexuality*, ed. Juliet Mitchell and Jacqueline Rose, 27–57. New York: Norton.

Rubin, K. H., and I. D. R. Brown. 1975. "A Life-span Look at Personal Perception and Its Relationship to Communicative Interaction." *Journal of Gerontology* 30: 461–68.

Rubinstein, Robert, Charles D. Laughlin, Jr., and John McManus. 1984. *Science as Cognitive Process: Toward an Empirical Philosophy of Science*. Philadelphia: University of Pennsylvania Press.

Ryan, E. B., and H. L. Capadona. 1978. "Age Perceptions and Evaluation Reactions toward Adult Speakers." *Journal of Gerontology* 33: 98–102.

Ryan, Michael. 1982. *Marxism and Deconstruction*. Baltimore: Johns Hopkins University Press.

Ryan, W. J. 1972. "Acoustic Aspects of the Aging Voice." *Journal of Gerontology* 27: 265–68.

Said, Edward. 1979. *Orientalism*. New York: Vintage.

———. 1983. *The World, the Text, and the Critic*. Cambridge, Mass.: Harvard University Press.

———. 1992. "The Politics of Knowledge." In *Debating P.C.*, ed. Paul Berman, 172–89. New York: Dell.

Salthouse, Timothy, Donald Kausler, and J. Scott Saults. 1988. "Utilization of Path-Analytic Procedures to Investigate the Role of Processing Resources in Cognitive Aging." *Psychology and Aging* 3: 152–66.

Sameroff, A. J. 1972. "Learning and Adaptation in Infancy: A Comparison of Models." In *Advances in Child Development and Behavior*, volume 7, ed. Reese, 169–214. New York: Academic Press.

Sampson, Geoffrey. 1980. *Schools of Linguistics*. Stanford: Stanford University Press.

Saussure, Ferdinand de. 1959. *Course in General Linguistics*. Trans. Wade Baskin. New York: McGraw-Hill.

Schacter, D. L., J. E. Eich, and Endel Tulving. 1978. "Richard Semon's Theory of Memory." *Journal of Verbal Learning and Verbal Behavior* 17: 721–43.

Schafer, Roy. 1980. "Narration in the Psychoanalytic Dialogue." *Critical Inquiry* 7: 29–53.

Schaie, K. W. 1977–78. "Toward a Stage Theory of Adult Cognitive Development." *International Journal of Aging and Human Development* 8: 129–38.

Schaie, K. W., and Gisela Labouvie-Vief. 1974. "Generational Versus Ontogenetic Components of Change in Adult Cognitive Behavior: A Fourteen-Year Cross-sequential Study." *Developmental Psychology* 10: 305–20.

Schaie, K. W., and C. R. Strother. 1968. "A Cross-sectional Study of Age Changes in Cognitive Behavior." *Psychological Bulletin* 70: 671–80.

Schallert, D. L., C. M. Kleinman, and A. D. Rubin. 1977. "Analysis of Differences between Oral and Written Language." ERIC Document Reproduction Service No. ED 146–418.

Schleifer, Ronald. 1980. "The Trap of the Imagination: The Gothic Tradition, Fiction, and 'The Turn of the Screw.'" *Criticism* 22: 297–319.

———. 1984. "Irony and the Literary Past: On *The Concept of Irony* and *The Mill on the Floss*." In *Kierkegaard and Literature*, ed. Ronald Schleifer and Robert Markley, 183–216. Norman: University of Oklahoma Press.

——. 1987a. *A. J. Greimas and the Nature of Meaning: Linguistics, Semiotics, and Discourse Theory.* Lincoln: University of Nebraska Press.

——. 1987b. "Deconstruction and Linguistic Analysis." *College English* 49: 381–95.

——. 1990. *Rhetoric and Death: The Language of Modernism and Postmodern Discourse Theory.* Urbana: University of Illinois Press.

——. 1991. "Analogy and Example: Heisenberg, Negation, and the Language of Quantum Mechanics." *Criticism* 33: 285–307.

Schleifer, Ronald, and Robert Markley. 1984. "Introduction: Writing without Authority and the Reading of Kierkegaard." In *Kierkegaard and Literature,* ed. Ronald Schleifer and Robert Markley, 3–22. Norman: University of Oklahoma Press.

Scott-Maxwell, Florida. 1979. *The Measure of My Days.* New York: Penguin.

Searle, John. 1979. "Metaphor." In *Metaphor and Thought,* ed. Anthony Ortony, New York: Cambridge.

Seligman, M. E. P. 1970. "On the Generality of the Laws of Learning." *Psychological Review* 77:406–18.

Sellars, Wilfrid. 1963. *Science, Perception and Reality.* London: Routledge and Kegan Paul.

Shelley, Percy. 1914. *Collected Poems.* London: Oxford University Press.

Sherman, N. C., and J. A. Gold. 1978–79. "Perceptions of Ideal and Typical Middle and Old Age." *International Journal of Aging and Human Development* 9: 67–72.

Sherman, N. C., and M. F. Sherman. 1978. "Attribution Theory and Evaluations of Older Men among College Students, Their Parents, and Grandparents." *Personality and Social Psychology Bulletin* 4: 440–42.

Shipp, T., and J. Hollien. 1969. "Perception of the Aging Male Voice." *Journal of Speech and Hearing Research* 12: 703–10.

Showalter, Elaine. 1985. "The Feminist Critical Revolution." In *The New Feminist Criticism: Essays on Women, Literature, and Theory,* ed. Elaine Showalter, 3–17. New York: Pantheon.

Simon, Eileen. 1979. "Depth and Elaboration of Processing in Relation to Age." *Journal of Experimental Psychology: Human Learning and Memory* 5: 115–24.

Spencer, H. 1916. *Principles of Sociology.* Volume 1. New York: Appleton.

Spicker, Stuart. 1978. "Gerontogenetic Mentation: Memory, Dementia, and Medicine in the Penultimate Years." In *Aging and the Elderly: Humanistic Perspectives in Gerontology,* ed. Stuart Spicker, Kathleen Woodward, and David Van Tassel, 153–80. Atlantic Highlands: Humanities Press.

Spivak, Gayatri Chakravorty. 1977. "The Letter as Cutting Edge." *Yale French Studies* 55/56: 208–26.

Steiner, George. 1971. *In Bluebeard's Castle.* New Haven: Yale University Press.

Stevens, Wallace. 1954. *Collected Poems.* New York: Knopf.

Stimpson, Catharine. 1982. "Editing *Signs.*" In *The Horizon of Literature,* ed. Paul Hernadi, 241–48. Lincoln: University of Nebraska Press.

Stimpson, Catherine, Joan Berstyn, Domna Stanton, Sandra Whisler. 1975. "Editorial." *Signs* 1: v–viii.

Stine, Elizabeth, and Arthur Wingfield. 1989. "Process and Strategy in Memory for Speech among Younger and Older Adults." *Psychology and Aging* 3: 272–79.

Stocking, George. 1968. *Race, Culture, and Evolution: Essays in the History of Anthropology*. New York: Free Press.

Strayer, David L., Christopher Wickens, and Rolf Braune. 1987. "Adult Age Differences in the Speech and Capacity of Information Processing: An Electrophysiological Approach." *Psychology and Aging* 2: 99–110.

Thomas, E. C., and K. Yamamoto. 1975. "Attitudes toward Age: An Exploration in School Age Children." *International Journal of Aging and Human Development* 6: 117–29.

Timiras, P. S. 1972. *Developmental Physiology and Aging*. New York: Macmillan.

Todorov, Tzvetan. 1975. *The Fanstastic: A Structural Approach to a Literary Genre*. Trans. Richard Howard. Ithaca: Cornell University Press.

Trubetzkoy, N. S. 1969. *Principles of Phonology*. Trans. Chrintinae Baltaxe. Berkeley and Los Angeles: University of California Press.

Turner, Barbara. 1979. "The Self Concepts of Older Women." *Research in Aging* 1:464–80.

Walsh, D., M. V. Williams, and C. K. Hertzog. 1979. "Age Related Differences in Two Stages of Central Perceptual Processes: The Effects of Short Duration Targets and Criterion Differences." *Journal of Gerontology* 34: 234–41.

Weber, Samuel. 1987. *Institution and Interpretation*. Minneapolis: University of Minnesota Press.

Weismann, A. 1891. *Essays upon Heredity and Kindred Biological Problems*. Clarendon: Oxford University Press.

Wheeler, Samuel C. 1989. "Metaphor According to Davidson and de Man." In *Redrawing the Lines: Analytic Philosophy, Deconstruction, and Literary Theory*, ed. Reed Way Dasenbrock, 116–39. Minneapolis: University of Minnesota Press.

White, Eric Charles. 1990. "Contemporary Cosmology and Narrative Theory." In *Literature and Science: Theory and Practice*, ed. Stuart Peterfreund, 91–111. Boston: Northeastern University Press.

Whitehead, Alfred North. 1967. *Science and the Modern World*. New York: Free Press.

Wilden, Anthony. 1968. *The Language of the Self*. New York: Delta.

Williams, G. C. 1957. "Pleitropy, Natural Selection, and the Evolution of Senescence." *Evolution* 11: 398–411.

Williams, Raymond. 1958. *Culture and Society*. New York: Oxford University Press.

Wilson, E. O. 1975. *Sociobiology: The New Synthesis*. Cambridge, Mass.: Harvard University Press.

Wolf, E., and A. M. Schraffa. 1964. "Relationship between Critical Flicker Frequency and Age in Flicker Perimetry." *Archives of Ophthalmology* 72: 832–43.

Wood, Vivian, and J. F. Robertson. 1976. "The Significance of Grandparenthood." In *Time, Roles, and Self in Old Age*, ed. Gubrium, 278–304. New York: Human Sciences Press.

Woodward, Kathleen. 1980. *At Last, the Real Distinguished Thing: The Late Poetry of Eliot, Pound, Stevens, and Williams*. Columbus: Ohio State University Press.

Woolgar, Steve. 1988. *Science: The Very Idea*. London: Tavistock.

Yeats, W. B. 1961. *Essays and Introductions*. New York: Collier.

——. 1971. *The Variorum Edition of the Poetry of W. B. Yeats*. Ed. Peter Allt and Russell K. Alspach. New York: Macmillan.

Young, Robert. 1981. "Editor's Introduction to Jeffrey Mehlman's Essay." In *Untying the Text: A Post-Structuralist Reader*, ed. Robert Young, 177–78. London: Routledge and Kegan Paul.

INDEX

Library of Congress Cataloging-in-Publication Data

Schleifer, Ronald.

Culture and cognition : the boundaries of literary and scientific inquiry / Ronald Schleifer, Robert Con Davis, Nancy Mergler.

p. cm.

Includes bibliographical references and index.

ISBN 0-8014-2632-4 (cloth). — ISBN 0-8014-9931-3 (paper)

1. Discourse analysis, Narrative. 2. Semiotics. 3. Cognition. 4. Psychoanalysis. 5. Science—Philosophy. 6. Communication and culture. 7. Criticism. I. Davis, Robert Con, 1948— II. Mergler, Nancy. III. Title.

P302.7.S37 1992

302.2—dc20 92-52770

Lightning Source UK Ltd.
Milton Keynes UK
UKHW040703200521
383975UK00012B/210